Continuous Intent Delivery

Continuous Intent Delivery

How Software Gets Built When AI Writes the Code

David Kim, Casey Robinson, Glenn Knepp

ISBN: 979-8-9925330-9-5 paperback
979-8-9960332-0-1 hardcover

Publisher: Self Published

Printed in the United States of America

First Edition

TABLE OF CONTENTS

DEDICATION

To the engineers, operators, and product builders who suspected the work had changed and were waiting for the language to catch up. The language is here. The argument is yours to test.

To our families, who watched the methodology emerge in real time and let us write the book.

EPIGRAPH

> "We make the road by walking."
>
> *Attributed to Antonio Machado, in Myles Horton and Paulo Freire, We Make the Road by Walking: Conversations on Education and Social Change (1990).*

PREFACE

This preface is written in three voices. David Kim, Casey Robinson, and Glenn Knepp are the three co-authors. Each speaks for himself in the section that bears his name.

I. David

I will spare the reader the writers'-room version of this book's origin and tell the operational version, because the operational version is the one that explains why the book exists.

Glenn Knepp and I met in 1996 at Marine Corps Base Hawaii, Kaneohe Bay. We lived in the same quadplex on base housing. I was a sergeant, active duty, just moving in with my wife and our newborn daughter. Glenn was a sergeant too, recently separated from active-duty. The day we met, I had locked my keys inside the unit. Glenn introduced himself, looked at the door, smiled, and walked away. A minute later he opened it from the inside. He had climbed up onto the back patio and slipped in through the sliding glass door. My first question was what's your MOS. He said Recon, of course. That began a friendship: years of building PCs out of our base-housing units, helping people and small businesses optimize their systems, then decades on separate tracks before we came back together to start Alchemaize. The Corps trains you to look at a system and ask whether it is going to fail under stress. Glenn and I have been asking each other that question for thirty years. The company we built runs on the answer.

The company is called Alchemaize. We founded it in August 2025 to ship an AI-augmented reading product called Ember. Ember was the plan, and the plan went the way most plans go in startups: it produced an MVP in December, missed its user-acquisition targets through a catalog problem we could not solve without a Kindle integration we had not yet been able to start a conversation about, and ended its first phase at the end of the year with the development team wound down and the company asking what came next.

In October 2025, Casey Robinson joined as COO. Casey and I have worked together on and off for almost twenty years, but the friendship is older than the working relationship. We met in 2006 in World of Warcraft, of all places. Casey raided alongside his wife, who I knew first; he and I bonded fast over wanting to play the game at the highest level the game had. We eventually left our guild together to start a new high-performance one, and that guild reached server-firsts and world-ranking raid clears. Out of that friendship came the first time I hired Casey: into a Texas state-government consulting engagement I was running. The work was Independent Verification and Validation under IEEE Standard 1012, against major systems-integration programs that ran north of one hundred million dollars each. TIER at the Texas Health and Human Services Commission. RTS at the Texas Department of Motor Vehicles. A few others. We spent five years walking into vendor work the state had paid for, validating that the requirements traced to the program's real intent, and verifying that the delivered software satisfied them. The structure of that work, validating intent on the front and verifying output on the back, with the team doing the validating and verifying held independent of the team doing the building, is recognizably the structure of CID. We did not know we were practicing for this.

By the second week of January 2026, Glenn, Casey, and I had concluded that the methodology Alchemaize had been operating under was not going to make Alchemaize's future. The AI generation tools we had been layering on top of a Scrum-shaped delivery process had spent the fall fighting that process rather than fitting it. On January 18 we drew a line in the sand, named the new approach Continuous Intent Delivery, and started over with the three of us as the pod.

Ten days later, on January 28, 2026, at four in the morning, Amazon issued a company-wide reduction in force. Sixteen thousand Amazonians received notices that their positions had been eliminated, with a 90-day notification period. My entire team of fourteen was hit, including Casey, who led a sub-team within my org, and including me. The methodology Casey had agreed to test ten days earlier became, in the space of one morning, the only work either of us had in front of us.

The 100 days between January 18 and April 20, 2026 are the case study at the heart of this book. Thirty-five applications shipped, across multiple domains, by three founders living in three different states, none of whom had ever shared an office. We had AI generation, written intent, executable verification, and the trust the three of us had built across decades of working together. We did not have standups, sprints, sprint reviews, retrospectives, PI Planning, or any of the other coordination scaffolding our industry has spent twenty years refining. The methodology did not need them, because what they were coordinating was no longer the bottleneck.

This book is the argument for what we found, the practice for how to do it, and the honest accounting of what it has cost the people who have tried to adopt it.

II. Casey

I want to say something about how I got here, because the version most people hear is shorter than the actual version, and the actual version is the one a reader of this book will probably find most useful.

I am not an engineer. I have never been an engineer. My first job in technology was at DirecTV, troubleshooting satellite systems for customers calling in from across the country. From DirecTV I went into consulting. The consulting I did, for five years alongside David, was Independent Verification and Validation under IEEE 1012. We were the firm the state hired to read everything before the state signed off on a vendor's work. Texas Health and Human Services on the TIER program. Texas DMV on RTS. Programs running north of one hundred million dollars each, with the state paying us to walk every requirement back to whether it actually said what the agency needed, and to watch the vendor deliverables to see whether they passed. Front of the pipeline, back of the pipeline. Validation and verification. Not building the system. Validating the intent it was supposed to carry, and verifying the system the build team handed back. I did not have a name for the structural shape of what I was practicing. The methodology landed in my hands and worked, mostly, because the front half had been my job already. From consulting I went into AWS, and from AWS I went into Alchemaize, and through all of those moves the role I played was operations. Run the planning. Talk to the customers. Translate what the customer wanted into language a delivery team could act on.

In October 2025 I joined Alchemaize as COO. The job description was ordinary. Run planning and operations for a small AI-driven startup, prepare for the Series A push off Ember. David and Glenn would build the product. I would build the runway around the product.

That is not what happened.

What happened is that in mid-January 2026, after Ember's first phase had ended, David and Glenn asked me to try something. They wanted to see whether the methodology they were starting to formalize, which they had just named CID, could be run by someone who was not an engineer. They picked a small project. The Trade Codex, a trading journal and education platform I had been thinking about on my own time, was the project I picked back. We agreed I'd try writing a VOS the next time we had a free video call. The free video call landed ten days later, after a Wednesday morning that converted David and me from part-time founders into full-time founders inside of a few hours, when Amazon's company-wide RIF on January 28 ended both our day jobs alongside the rest of our team's. I tell that morning's story properly in Chapter 8. What it meant for the methodology is that the Thursday after, January 29, when David finally shared his screen and walked me through opening my IDE, the experiment we had been planning had become the only thing either of us had to do.

The document was a few pages long. It said what the product was, who it served, what the user would see when they used it, what files the AI generator needed, and what would be measurably true if the product worked. It was a VOS, the unit the methodology had named ten days earlier. I sent it to Glenn for review. He came back with feedback on three of the four acceptance contracts. Two of his rejections were correct. On the third, my perspective was stronger and we talked it through. The fourth he accepted. The next morning we ran the generation, and twenty minutes later I had a working application on my screen.

A smile crept across my face in the middle of the night, just amazed at what just occurred. I sat there for a while. Then I started writing the next one.

I want to say something about what that moment was, because it is the moment a reader is going to want to argue with. The moment was not magic. The moment was the result of David and Glenn doing fifteen years of architectural work that I got to walk into. The methodology was the gift they handed me. What I did was use it. The point of this book is that the methodology is now also a gift you can walk into, and the question the book is asking is whether you, the reader, are willing to use it.

III. Glenn

The methodology in this book is older than the book. We named it in early 2026. The discipline behind it goes back further. I want to explain what discipline.

I served in the Marine Corps from 1988 to 1996, including time as a Signals Intelligence Specialist and a Special Operations Reconnaissance Team Leader. After the uniform came off, I spent years on the civilian side of the U.S. intelligence community, building and verifying classified systems whose failure modes were not theoretical. From there I moved into enterprise platform work, into startup founding, into a portfolio of patents on cloud cryptography, and eventually into Alchemaize with David, who I have known since the year I left active-duty.

What carried through every one of those environments was a single instinct. Before you ship a system, you write down what the system has to do. Then you verify, before you ship, that it does it. Verification is not a phase that comes after. Verification is a contract that comes first. Without that contract, the system is a hope. With it, the system is a claim that can be tested.

That is not novel. It is the entire history of formal methods, the entire history of safety-critical engineering, the entire history of the regulated environments I have worked in. What is new is that the cost of generating the system from the contract has collapsed. Twenty years ago, writing the contract first cost a team months of engineering discipline they could rarely afford. Today, the contract is the only durable artifact, and the engineering discipline is the cheap part. The methodology in this book is what falls out when you reorganize a software organization around that fact.

I will be direct about what that reorganization costs. It costs the rooms, the ceremonies, and the feeling of shipping software through a sequence of agreed-upon meetings. It costs senior engineers their identity in the act of typing, which has been the central skill of our industry for half a century, and which is now an auxiliary one. The book names every one of those costs, because a methodology that hides its costs is a methodology that fails on contact with a real organization.

What the reorganization buys is straightforward. It buys software that ships at AI speed without the failure rates that AI speed produces under the old methodology. It buys verification rigor at the start of the pipeline rather than as a final approval. It buys a team that knows what it is shipping, and a finance function that knows what it is funding, and a board that can read the four numbers that actually run the engineering organization.

The argument is the rest of the book. Read it. Push back on it where you should. We have tried to be honest about what we know and where we are still learning.

IV. David, briefly

We considered writing this as a white paper. The argument fits in a white paper. We did not write a white paper because the argument is not, in the end, a technical argument. It is an argument about how organizations will spend their attention in the next ten years, and that argument is one a reader has to walk through, not skim. A white paper produces nods. A book produces decisions.

What we are asking the reader to do is not to adopt our methodology because we have. We are asking the reader to read carefully enough to decide whether the argument is right, and if it is, to begin a small experiment in their own organization to test it. Chapter 13 names the experiment. Appendix C is the readiness assessment. The path is short and the first step is small. We hope you take it.

David, Casey, and Glenn April 2026

PART I

The Argument

Chapters 1 through 3.

The bottleneck moved. The frameworks that coordinated the old bottleneck have become friction. The evidence is one team's hundred days, with the counting laid down flat.

CHAPTER 1: THE SHIFT

In February of 2026, in home office on the ground floor of a house in Claremore, Oklahoma, Casey deployed his forty-second Verifiable Outcome Slice of the quarter. He was two states away from Glenn, five hundred miles from me, and not in a shared building with either of us. He had been at his desk for about five hours. The house was quiet. The code that shipped, across three files in an application called The Trade Codex, was going to serve real users the following morning.

Casey is the third author of this book. He wasn't a software engineer; in any sense the industry would have recognized in 2025. His career is customer service, operations, consulting, and Agile program management, carried over two decades. For four years immediately before Alchemaize he was at Amazon Web Services at the Senior Manager level, leading a team of Customer Solutions Managers, the people who guide enterprise customers through cloud migrations. He had never written a line of production code in his working life. He was hired as COO in October 2025 to run planning and operations and to get the company ready for a Series A raise we were anticipating off our reading application, Ember. He wasn't hired to ship software. Nobody expected him to.

Then the first 100 days didn't work, and the next 100 days had to.

In the hundred days between January 18 and April 20, 2026, Casey shipped dozens of VOSes against production code across four products, TradeCodex, Finaize, SkipDay, and BidForge, without Glenn or me touching the code that came out of his work. That fact is one of the three reasons this book exists. The other two are why those hundred days happened at all.

This chapter is about what had to be true for that fact to be possible.

1. The first 100 days, and the day of reckoning

I want to say what happened before the second 100 days began, because the methodology we are about to describe was not designed at a whiteboard. It was forced into existence by a specific stretch of work that did not work, and by a morning ten days into the new approach that converted three part-time founders into three full-time founders inside of a few hours. The book has been working for some time on the soft version of this story. The honest version is below.

Alchemaize was incorporated in August 2025. Glenn and I have known each other since 1996, when we met in a Marine Corps base-housing quadplex at MCBH Kaneohe Bay, Hawaii, and the company has from the start carried that texture: we trust each other, we say what we mean, and we build the thing in front of us until either it works or it doesn't. Casey joined as COO on October 1, 2025. By October 20 we had a development team in place. Three developers and an AWS solutions architect, all part-time, all comped on options and deferred contingent compensation tied to seed and Series A milestones, all holding full-time jobs at FAANG companies on the side. The plan was to build Ember the way the industry knew how to build software. Standups in the morning. Sprints. Sprint planning. Retros. Sprint reviews. AI tools layered on top of the methodology, used as developer-productivity multipliers inside an otherwise familiar pipeline.

That plan ran for the rest of the fall. The Ember web-app MVP shipped on December 1, 2025. The MVP worked. What it could not do, on the catalog the audience actually wanted, was reach those readers without a Kindle integration we could not yet open a conversation about. Ember worked beautifully on Project Gutenberg. The target audience wanted Harry Potter and Fourth Wing, and DRM made the commercial-title catalog inaccessible to us as long as we were outside the Kindle ecosystem. The user-acquisition story we needed to tell to a Series A round depended on the catalog. The catalog depended on the integration. The integration depended on a relationship we had not yet been able to start.

By the end of December the answer was clear. Ember was a product Alchemaize had built well and could not yet sell at the scale a venture round required. The development team and the architect were wound down. Because the team had been on contingent comp tied to milestones we were not going to hit on the original schedule, no severance was owed; the team returned to their full-time roles. We did not let people go from livelihoods. We released them from a side project that had run its course.

January 9, 2026, was day 100 of Alchemaize operating in the old way. The first 100 days had produced a finished MVP, a four-person team that came and went, and an answer to a question we had not realized we were asking. The answer was that the methodology we had been operating under was not going to make Alchemaize's future. The same AI tooling that had been making our individual work faster was, inside a Scrum-shaped delivery process, fighting the methodology rather than fitting it. Two ways of working at war over how a piece of work was characterized, executed, and released.

By the second week of January, the three of us had a conclusion. Alchemaize had to fundamentally change how it worked or it did not have a future. The seeds of CID and ELCID were planted in the conversations during that stretch. On January 18, 2026, we drew a line in the sand. The methodology had a name. The pod was the three founders. No development team, no architect. Three people, three states, three home offices. We were going to operate the new pattern across every project we could touch, and we were going to find out whether it produced something Ember-the-old-way had not.

Ten days later, on January 28, 2026, at 4:00 AM Central time, Amazon issued a company-wide reduction in force. Sixteen thousand Amazonians received notices that their positions had been eliminated. The notification period ran from January 28 through April 28, 2026. No work expected, full pay and benefits for ninety days, separation date of April 28. My entire team of fourteen Technical Customer Solutions Managers, Business Development Managers, and Solutions Architects was hit. Casey, who led the early-migrations and EBA sub-team within my broader org, was hit. I was hit.

I realized what had happened around six. I tried to access an Amazon resource and could not. I texted Casey at 6:05 AM. *Hey, I think our whole team got hit including me.* Casey was asleep. He woke at 8:00 to a wall of texts on his phone. Some of them were from his own team members, every one written in the same shape. *Hey Casey, I was impacted by the RIF, hope you are okay, it was awesome working with you.* They all assumed Casey, as their manager, was safe. He was not. He texted me back at 8:18. *Yep...* By 8:20 we were on a Zoom call together, working through what to say to the team and what kind of support we could offer the people whose mornings looked the way ours did.

What we landed on, that day and across the next several weeks, was a Discord channel for the full fourteen-person team. A place to communicate, to vent, to ask each other for leads, to be present for each other while everyone figured out what came next. The channel is still active. People handle these mornings differently, and the channel made room for that.

The layoff did two things to Alchemaize. The first is structural and the book will name it bluntly. It converted Casey and me from part-time founders, working around AWS day jobs, into full-time founders, with ninety days of paychecks already arriving and no day jobs to return to. The second is harder to name. The first 100 days had ended with us asking whether Alchemaize had a future at all even with the new methodology, and the layoff arrived in the middle of that question and rearranged it. We were not going back to AWS. The methodology we had drawn the line in the sand for ten days earlier was now the work in front of us, full-time, with the paychecks counting down to April 28. The book is honest about both halves of what that morning was. It was a blow. It was also a forcing function. Both at once.

What the next ninety days produced is the case study at the heart of this book. By April 20, the second day 100, the pod had shipped thirty-five production applications, roughly one million lines of production source code, 429 test files, and two thousand commits across the portfolio. Three founders. Three states. No shared office. The methodology that produced those numbers, refined as we went, is what the rest of this book describes.

The chapter begins again, on argument, below.

It is also a book about the thing that fact implies, which is that a twenty-five-year consensus about how software gets built is over. We'll spend the next 80,000 words making that case. First as an argument, then as a methodology, then as a set of enterprise practices, and last in a chapter Casey wrote himself, as a lived experience. By the end we'll have done something the current literature on AI and software development has not yet done. We'll have described what replaces Scrum.

I want to say that plainly before we go further, because a lot of what follows depends on it. This book is not *AI and Agile, together at last.* It is not *how to add prompt engineering to your sprint.* It is not one of the many books being written right now about how to use AI coding tools inside an otherwise unchanged engineering practice. Those books have their place. This is not that book.

The argument in your hands is that the engineering practice itself has to change, because the constraint it was designed for is gone, and the constraint it now faces is not one any existing framework was built to hold.

That sentence is the whole claim. The case for it takes a chapter. What to do about it takes the rest of the book.

2. The bottleneck moved

For twenty-five years, every method for building software at scale has been organized around a single assumption. The assumption was that humans are the bottleneck of the implementation step, that typing code is the expensive, slow, error-prone activity, and that the job of a framework is to coordinate humans around that expensive activity so their typing-output gets used well.

Every framework you know, descended from every framework that preceded it, is a solution to this problem. Waterfall said plan exhaustively up front, because the typing is the long pole and mistakes are expensive to rewrite. Scrum sliced the work into two-week increments, because the typing was still the long pole but requirements changed faster than the plan. Extreme Programming paired the typists so they caught each other's errors in real time. Kanban limited concurrent typing so throughput didn't collapse under context-switching. SAFe coordinated large groups of typists at ten-week resolution so the organization could plan. DevOps automated everything around the typing that wasn't typing, so the typists could concentrate on the typing. Continuous Delivery declared that the typing was fine; what happened after the typing was broken.

Every one of those frameworks is a rational, careful, well-engineered answer to a single question. Typing is expensive. How do we organize an organization around that fact?

The question has changed.

It's not that AI has made typing free, though in some meaningful sense it has. It's that the step between the human specification and the running code, the step we used to call implementation or coding or, on a bad day, cutting the ticket, has become, in the hands of a competent operator, roughly the cheapest step in the pipeline. Cheapest is a specific claim. It isn't a claim about clock time (some generation runs are slow). It's a claim about where the constraint now sits: somewhere else entirely. The step isn't worth organizing around.

It isn't precisely free, of course. There are still costs: model-inference costs, verification costs, operator costs, the real cost of the time it takes to write the specification well. But those costs aren't typing costs. They're the costs of the other steps, steps that used to be cheap relative to typing because typing dominated them. When the dominant cost is removed from a system, the structure of the system does not stay the same. It shifts, all at once, to fit the new dominant cost.

The new dominant cost is two things, not one. They sit on either side of the old bottleneck.

The first is *intent.* Specifying, unambiguously and completely enough that a machine can act on it, what you actually want the software to do. In the old world we called this requirements engineering and treated it as a ten-percent activity because typing was the ninety-percent. In the new world, intent clarity is the dominant input to output quality. If you're any good at this work now, you've already noticed it. The difference between an AI tool that does what you wanted and an AI tool that does some plausible thing you did not want is almost always a difference in the specification going in. There's no prompt-engineering trick that compensates for intent that was vague from the start.

The second is *verification.* Knowing, with reasonable confidence, that the code you shipped does what you intended it to do. In the old world we called this quality assurance and treated it as a downstream activity because typing was the expensive middle. In the new world, verification is the structural guarantee of the whole pipeline. If you cannot verify, you cannot ship at any speed, because shipping unverified AI output is a distinct and self-amplifying form of disaster, one this book will name specifically in a later chapter.

The bottleneck didn't go away. It moved. The typing step got cheap. The intent step and the verification step got dominant.

There's a parable in Eliyahu Goldratt's *The Goal* [2] about Herbie, the slowest scout in a boy scout troop on a hike. Herbie sets the troop's pace; the faster scouts in front of him cannot pull away because the line has to stay together. Goldratt built the Theory of Constraints around what to do about Herbie: identify the bottleneck, elevate it, and watch throughput rise until the next constraint shows up. Half a generation of operations management runs on that book. What AI generation did to software is not the move Goldratt's framework predicted. We did not elevate the typing bottleneck. We removed it. The Herbie of the software pipeline got handed a jetpack, and the troop is now governed by two constraints that used to be almost invisible from the back of the line. Intent at the front. Verification at the back. Goldratt's lesson still holds. The bottleneck always exists, just not where it was.

Pipeline step	Before AI generation	After AI generation
Intent (what to build)	Existed. Treated as a 10 percent activity because typing was 90.	Existed. Now the dominant cost at the front of the pipeline.
Verification (confirming what shipped)	Existed. Treated as downstream QA, a safety net under the typing.	Existed. Now the dominant cost at the back of the pipeline.
Typing (turning spec into running code)	Dominant. Ninety percent of the work.	Cheapest step. In an operator's hands, a rounding error.
Organizations	Built around typing.	Still built around typing. This is the problem.

Table T1 · What changed, what didn't.

Three of the four rows haven't moved. Intent was always there. Verification was always there. Organizations are still there, coordinating the same thing they've always coordinated. The fourth row, typing, is the one that moved. Everything structural follows from the mismatch between rows one through three and row four.

The implication, if you let it land, is that every framework currently in use at your company was designed for a bottleneck that no longer exists. That's a structural observation, not a polemical one. The thing Scrum's machinery is best at, coordinating the cost of human typing across a sprint, is no longer the cost that matters. The thing SAFe's machinery is best at, coordinating hundreds of human typists at ten-week resolution, is coordinating something that doesn't need to be coordinated anymore. Not as a criticism of the typists. As a statement about who's typing, which is now the machine.

The frameworks have become friction.

That's the structural observation of this book. Everything that follows, the methodology, the roles, the metrics, the adoption path, is an attempt to answer the question that opens up the moment you accept that observation. What, then, is the thing we should be building our organizations around?

3. A concession before we continue

The sentence "the frameworks have become friction" can sound glib, and I've been in this industry long enough to know what happens when a new methodology talks about the old one that way. A third of readers stop reading, because they've heard this kind of thing three times this year already. Another third reads defensively. The last third reads eagerly, waiting for permission to say what they already thought. I don't want any of those three readers right now. I want the reader who's willing to judge on its merits.

So let me be fair to the frameworks we're about to name. Scrum was a reasonable solution to a real problem. SAFe was a reasonable scaling of that solution to larger organizations. The individual practices these frameworks advocated, the standup, the retro, the sprint review, the Program Increment, were not absurd. *The Scrum Guide* still describes the Daily Scrum as a 15-minute event for the developers to inspect and adapt the sprint plan [1], and that description, read on its own terms, is sensible. These were, and in many contexts still are, the best answer a careful engineering culture ever gave to the question of how to organize humans whose typing was the long pole.

The criticism isn't that the frameworks were wrong then. The criticism is that the condition they were designed for is no longer the condition we're in. A design rational for one constraint becomes an impediment under a different constraint. That's how engineering works. That is, for that matter, how software works.

So when the next chapter takes a scalpel to the specific thirty-two hours of ceremony a Scrum team currently pays per sprint and demonstrates that the same coordination function can be accomplished in thirty minutes per week under the new constraint, please read that chapter the way I intend it. Not as a criticism of Scrum the framework, but as a description of the structural cost of running a machine calibrated for one constraint in a world that has moved to another.

4. What this is not

There are three books this is sometimes mistaken for. We should put those mistakes to bed.

This isn't a book about AI tools. It doesn't endorse a particular IDE plugin, a particular model, a particular API, or a particular way of prompting. It won't care whether you use Claude or GPT or whatever comes next. The methodology we'll describe works with any competent generation layer, and the specific generation layer is the cheapest and most replaceable part of the system it sits in. The book's argument would survive a complete turnover of every AI coding tool currently on the market. What it depends on is the existence of *a* generation layer capable of producing verifiable code against a specification. That layer exists in several forms now and will exist in more forms a year from now. We're writing past the specific tools to the condition they collectively create.

This isn't a book about AI replacing software engineers. The methodology we'll describe has three roles at the team layer and three more at the enterprise layer. Every one of those roles is held by a human. Several are held by humans who have been career software engineers for decades. What has changed is not whether humans are in the loop. What has changed is what the humans are doing in the loop. The expensive, rare, specifically skilled work has shifted from typing correct code to specifying intent clearly, curating context precisely, and verifying output rigorously. The good software engineers will still be rare and specifically skilled. They'll just be rare and specifically skilled at something else.

This isn't a book about prompt engineering. It doesn't contain tips on how to phrase a prompt to get a better completion. If you want that, there are fine books; this isn't one of them. Prompt engineering lives inside a much larger question. What specification is a human supposed to be authoring? What contract is the verifier supposed to enforce? What context is the system supposed to be operating on? What outcome is the slice of work supposed to produce? Prompt engineering is the smallest layer of the stack the book concerns itself with, and we won't be discussing it at length. The work in this book lives at the layers above it.

5. The new constraint and what it demands

If the bottleneck has moved to intent and verification, then a method for building software at scale under the new constraint needs to be organized around those two things. That's the whole design brief. The rest is details.

One useful way to see what this looks like in practice is to notice what has not changed.

Business outcomes haven't changed. You still need to know what you're trying to build and why. What has changed is that you now need to write that down with more precision than you used to, because something is going to act on it without the soft rails of a twenty-year-old engineering team's shared context. The intent specification is no longer a communication artifact between humans who already know most of the answer. It's a real input to an actor that does not know the answer and will act on whatever you give it.

Users haven't changed. You still need to verify that what you built works for them. What has changed is that you now need to verify more, and earlier, and continuously, because the output is arriving faster than any QA process designed for human-speed generation can keep up with. The verification step hasn't been bypassed. It's been inverted. Brought upstream of generation, made continuous, made the structural guarantee rather than the downstream check.

Organizations haven't changed. You still need to decide what you're going to fund and why. What has changed is that the unit of funding, the project, with its fixed scope and schedule and deliverable, was an instrument for a world in which humans were the constraint and scope was negotiable. The new constraint asks for a different instrument.

The methodology you'll meet in the next chapters takes those three observations and builds from them. It's called Continuous Intent Delivery. CID at the team layer. ELCID at the enterprise layer. The chapters to come will explain what that means, how it works, what it costs, and how you adopt it. But the central design move is already visible. The methodology is organized around the things that have become dominant: intent clarity, verification rigor, and outcome-based funding. It's not organized around the thing that became cheap, which is typing.

The shortest way to say this, and the book will earn the right to shorthand only sparingly: *intent is the artifact, code is the exhaust.*

I need to be precise about what that means, because the sentence does a lot of work and will show up in later chapters more than once. It does not mean code is unimportant. Of course code is important; it's what runs. It means something specific. The artifact that humans should author, version, review, argue over, maintain, and treat as the organization's durable knowledge is the *specification of intent*, not the generated implementation. The code is what comes out the other end when the specification is right. The code is, in a reasonable sense, the predictable output of a well-run pipeline, not the creative output of a human act.

A reader with the instincts of a career engineer will find this formulation uncomfortable. I know. I was a career engineer. The discomfort is the right response to the sentence. What the sentence is doing is naming the inversion, and the inversion is uncomfortable because it reorganizes where the creative act lives. It doesn't abolish the creative act; it relocates it. Chapter 4 relocates it carefully.

6. The proof that the inversion holds

There is a headline number associated with the methodology we're describing. It's a real number, and it's the kind of number that makes a person pay attention. A pair of adjective-free facts that describe an outcome that sounds, on first hearing, implausible. You can read it in Chapter 3; the chapter we wrote to hold it. Chapter 3 shows what we counted, what we didn't count, which projects we killed and why, and what the measurement methodology was. We're holding the number to Chapter 3 on purpose. Introduced in Chapter 1 it becomes a sales pitch. Introduced in Chapter 3, with everything around it disclosed, it becomes evidence.

The evidence is the load the rest of the book rests on. I'm asking you to wait for it.

What I'll tell you in Chapter 1 is two things about the evidence, without yet giving you the numbers.

First, the evidence isn't that we built software faster. Faster is too small a claim, and it's the claim every framework ever written has made about itself. The evidence is that we built *different* software, in a different *shape*, by a different *set of people*, than any of us could have built under the old constraint. The shape of the output is evidence in a way that velocity metrics are not, because shape is harder to fake and harder to mistake.

Second, the evidence includes Casey. The third author of this book is not a software engineer. He hadn't written a unit test in his professional life before the day in January of 2026 when I asked him to try writing a specification instead of a wish list. He shipped production code during the period the evidence covers. Not two lines of copy-paste. Features. Full features, with acceptance contracts and verification passes and real users, in an application that processes financial documents for small businesses.

That particular fact, when I first encountered it as a claim about what the methodology was capable of, did more to convince me the methodology was real than any productivity number would have. It also wasn't an accident. Casey and I had spent five years before our AWS work doing Independent

Verification and Validation under IEEE 1012, against major Texas state systems-integration programs. The job was to read every requirement on the customer side, validate that it said what the agency actually meant, and watch the vendor's deliverables to see whether they satisfied the requirement. Front of the pipeline, back of the pipeline. Validation and verification, held independent of the team doing the building. Twenty years on, the methodology that fell out of CID asked for the same skill, named differently. Casey came in pre-trained for the Intent Engineer role. He just didn't have a name for the role yet. Productivity numbers come and go, and most of them are inflation hiding in decimal points. A non-engineer shipping production code against a verification gate is not something a spreadsheet can hide. It's either true or false. It's true. Chapter 8 describes how, in Casey's own voice.

7. A working definition

We're going to go slowly with the methodology. Chapter 4 walks through the atomic unit of work, the VOS, for Verifiable Outcome Slice, in enough detail that you can draft one by the end of the chapter. Chapter 7 walks through the pipeline a VOS travels and the watching layer that runs alongside it. Chapter 6 walks through the verification step, which is the structural thing that makes the whole pipeline work at speed. Chapter 9 walks through the funding model that makes it scale.

A note on geometry. CID is a forward pipeline per VOS, with two reverse edges that exist to correct fast when verification surfaces a gap. SHIPPED is terminal. Observation watches deployed VOSes in production, in parallel, and feeds new intents into the head of the stream. The shape is not a circle. It is closer to a Toyota line than to PDCA, OODA, or Scrum. This is also not waterfall. The cycle time per VOS is hours or days, not months. The reverse edges allow rapid contract-and-regenerate iteration without big-design-up-front. Observation runs in parallel rather than as a final QA phase. Pipeline geometry plus eddies plus a watching layer is its own thing. The rest of the book builds it up, piece by piece, until the picture is whole.

You should leave this chapter with a working definition, because the argument of the rest of the book depends on having the shape in mind.

Continuous Intent Delivery is the practice of building software by:

1. Authoring verifiable intent specifications: small, self-contained slices of work, each with a written outcome it's intended to produce and an executable contract that will confirm the outcome was produced.
2. Delegating implementation to an AI generation layer: any competent model capable of producing code against a contract and a curated context bundle.
3. Moving verification upstream of generation: writing the acceptance contract before the code, enforcing it during the code's production, running it continuously against the code after it ships.
4. Organizing a small, role-structured team: three roles, three humans or fewer, around the pipeline that carries these three activities.

That's CID at the team level. The enterprise layer, ELCID, for Enterprise Layer for Continuous Intent Delivery, scales this by replacing project funding with intent streams, replacing the PI Planning cycle with a ninety-minute monthly review, and replacing the fourteen-role coordination structure of SAFe with a six-role structure that sits on top of the team-level pods.

You can hold all of that in your head. That was the point of the design. The *documentation* of the methodology, the book you're holding, is longer than the methodology itself, because methodologies need arguments and examples and diagnostic instruments, and the book supplies those. The methodology itself, the thing a team actually runs, is small.

The smallness isn't accidental. A methodology that's too big to hold is a methodology that accretes ceremony. We named that specific failure mode and gave it a name so readers would recognize it in their own organization, in Chapter 12. It's called *Ceremony Creep*. It's a predictable failure of this framework, just as it has been a predictable failure of every framework that preceded it. We won't solve it by wishing. We'll solve it by naming it.

8. A note from the Verification Owner

A short interruption from one of your other co-authors. He shows up properly in Chapter 6 but wants to put one thing on the record now.

The whole argument you've just read, the bottleneck has moved, typing got cheap, intent and verification got dominant, has a failure mode. A current one. It's happening right now in every engineering organization adopting AI tools without adjusting its methodology. The failure mode is simple. The downstream verification machinery that was designed to keep up with human typing cannot keep up with machine typing. The organization doesn't notice the verification has failed until a customer finds a bug that a test would have caught and that there was no test for.

This isn't theoretical. It's happening in teams you've heard of. I've watched it happen. It will happen in this book's first quarter of readership, and some of you will call me later and tell me it happened to you.

The rest of this book has one structural concern, carried like a bass note under the methodology: do not ship AI-speed generation into a human-speed verification setup. The methodology you're about to learn gives verification equal architectural standing with generation. That's the only design choice that prevents the failure mode above. Everything else, the streams, the roles, the metrics, the anti-patterns, is built on top of that one design choice.

I'm setting this down here, in Chapter 1, so that when you reach Chapter 6 and find that the Verification Owner is not a QA manager, does not report to the delivery chain, and owns something called an *acceptance contract* that looks alien the first time you see it, you'll know why. The thing the Verification Owner is protecting is the only thing standing between AI-speed generation and AI-speed disaster.

Back to you, David.

---

9. Who this book is for

I want to be explicit about the reader this book imagines, because the reader changes as the chapters progress and I want you to recognize when the book is speaking to you specifically.

If you're a CTO, VP of Engineering, or Head of Platform, the book is speaking to you through the argumentative spine of the chapters, and particularly through Part III, which is the ELCID enterprise layer. Chapter 9 is about how to change your funding model. Chapter 10 is about how to change your metrics. Chapter 11 is about how to do both in a regulated industry. Chapter 13 is about how to adopt any of this without launching a transformation program.

If you're a senior engineer or engineering manager, the book is speaking to you through the practice chapters in Part II. Chapter 4 will restructure your intuitions about what "specification" means. Chapter 5 will change your practice around what you feed a generation tool. Chapter 6 will change your practice around how you verify. You're the reader most likely to recognize, by the end of Part II, that there is a version of your career you're about to begin.

If you're a non-engineer, an operator, a founder without a technical co-founder, a domain expert who has always had to route software requirements through someone else, the book is speaking to you most directly through Chapter 8. That chapter is Casey's. It is written by one of you. You may find you can read it before you read anything else, and if you do, the rest of the book will retrofit into the picture that chapter gives you. Some readers will reach that chapter and realize they've been waiting for a methodology like this their whole working life. We wrote the chapter for you.

The book isn't trying to be all things to all readers. It's a methodology book. But the methodology has enough explanatory power, and enough cross-functional reach, that different readers will get different books out of it. That's a feature of what the methodology is. It changes what a cross-functional team is, and we designed the book to match.

10. The road ahead

The rest of this book proceeds in four parts.

Part I, which this chapter opens and Chapters 2 and 3 complete, is the argument itself. Chapter 2 confronts the existing frameworks, specifically and by name, and shows that the coordination costs they impose are not marginal tuning problems but structural impediments under the new constraint. Chapter 3 is the evidence chapter, where we lay out what we built in one hundred days and show our counting methodology.

Part II, Chapters 4 through 8, is the practice. This is where the methodology actually lives. Chapter 4 defines the atomic unit and introduces the pipeline. Chapter 5 is on context curation, a skill that separates a five-times productivity multiplier from a thirty-times one, written in Casey's voice because context curation is Casey's domain. Chapter 6 is on verification, Glenn's chapter, and the chapter some readers will end up highlighting more than any other. Chapter 7 is on the pipeline in motion, the weekly and daily mechanics of running CID inside a pod. Chapter 8 is Casey's showcase, the *Phoenix Project* chapter of this book, except that Casey is real, he's one of the authors, and the events described are things that actually happened.

Part III, Chapters 9 through 11, is the enterprise layer. Chapter 9 is about funding: replacing projects with intent streams. Chapter 10 is about metrics: replacing the velocity-and-predictability dashboard with four numbers. Chapter 11 is about compliance, a chapter that will be especially relevant if you work in healthcare, finance, or government, and that argues a position some readers will find surprising: the regulated industries are the *strongest* candidates for AI-native development, not the weakest.

Part IV, Chapters 12 through 14, is the road. Chapter 12 is the anti-patterns, the specific ways this methodology fails when it fails. Chapter 13 is the adoption path, deliberately pragmatic and deliberately not a transformation program. Chapter 14 is the closing chapter. It attempts to answer a question this introduction has deferred: *who, exactly, gets to build* *software in the world this methodology describes?* We'll answer it with two voices alternating, and Casey's voice will close the book.

A methodology book with three authors has to be careful about whose voice is in what chapter. We've been careful. The three of us did not blend into a single corporate-first-person "we." Each chapter declares its primary author in the opening lines. Casey's running sidebars, *Field Notes from the Test Pilot*, are visually distinct and appear in every chapter except the one Casey wrote by himself. The book has one editorial pulse, but three voices inside it, and the texture of the prose changes when the voice changes. That's the right way to write a book whose thesis is that building software should stop being one kind of work done by one kind of person. A book with one voice would be failing its own argument.

The chapter should close with the voice of the person the methodology was actually for. Casey will tell you, in a short Field Note, what he was hired to do at Alchemaize and what he ended up doing instead.

Field Note from the Test Pilot

> *Casey Robinson, Intent Engineer*
>
> **What I was hired to do, and what I ended up doing instead**
>
> Before I get into the work, a quick bit of context about me. I'm from Claremore, Oklahoma, and a big sports guy, which means the way I look (six-two, big frame) usually gets me read as a retired football player or a football coach before anyone reads me as a consultant or a technologist. That's the setup, and it's relevant, because the thing I do for a living now is author the specifications for production software that runs against real customer data. If you'd told me, on the day I was brought into Alchemaize, that that was going to be part of my job, I'd have laughed.

I was brought into Alchemaize to run operations for a three-person startup building an AI-augmented reading application called Ember. That was the job description I signed, and it felt like the right shape for me, because my background is not in software engineering or anything related to building a codebase or deploying infrastructure. My career is customer service, consulting, operations, and Agile program management, carried for long enough that it's the shape my working life has. For four years immediately before Alchemaize I was at Amazon Web Services, at the Senior Manager level, leading a team of Customer Solutions Managers, the people who stand in the seam between an enterprise customer and the technical capability they're trying to adopt. You guide the migration. You hold the plan together. You translate between the customer's internal politics and the technology's actual shape. You do a great deal of it in Agile ceremonies, because that's the format enterprise delivery has mostly settled into. You don't write production code. That was my world, and variations of it have been my world for most of my working life. The under-the-hood parts of how software actually runs are, mostly, a foreign language to me. David and Glenn know the code. I know what a customer-steering committee needs to hear, how to read a room of investors, why the company insurance is due on the seventeenth and not the first. That was the arrangement. We'd never shared an office either. Alchemaize is three founders in three states: Brentwood, Tennessee; Claremore, Oklahoma; College Station, Texas. A catalogue of home-office rooms that used to be dens, spare bedrooms, and studies. No building. There was never going to be a building.

The arrangement held through the fall and into the new year. Ember's MVP shipped December 1. The reading product worked beautifully on public-domain titles and could not, without a Kindle integration we could not yet open a conversation about, get to the catalog of commercial fiction the target readers actually wanted. The user-acquisition numbers reflected the catalog problem. By the end of December we'd wound the development team down. The Series A push I'd been hired to run was, by early January, not the path forward.

Then in early January, David and Glenn told me they thought there was something different to try. They'd spent the fall watching their own work get faster in a way the methodology they were inside of could not absorb, and by the second week of January we'd agreed Alchemaize was going to operate a different pattern. They had a claim they couldn't yet verify. The claim was that a person who was good at being clear about what a business needs software to do, which was my work, could, under the right framework, ship production software directly, without handing the work off to an engineer in the middle. I was the non-engineer on the team. They asked me if I'd be willing to test the claim, and they were clear that if it didn't work out we'd all learn something and move on. We drew a line in the sand on January 18 and started.

Ten days later, on January 28, the floor came up to meet us. David and I were both at AWS at the time, alongside a fourteen-person team I'd helped lead the migrations sub-team inside of, and at four in the morning a company-wide RIF notice went out. David and I were on it, along with the entire team. By 8:20 the morning of, David and I were on a Zoom call together, trying to be useful to people whose mornings looked the same way ours did. I tell that story properly in Chapter 8. What it meant for the methodology is that the experiment David and Glenn and I had agreed to ten days earlier became, in the space of one morning, the work. There was no Series A push to go back to. There was no AWS day job to fail back into. The methodology they were asking me to test was now the runway, and the runway was the next ninety days.

I was curious, I wanted to help, and I'd spent enough years at AWS watching enterprise customers struggle with the same specification-and-handoff problem that I already had a theory the handoff was the expensive step. I assumed I'd probably fail, and we'd figure out what came next. I was wrong about the first part of that assumption, and the second part turned out to be true in ways none of us expected.

There wasn't a moment of clarity. What there was, was a few weeks of writing small, specific descriptions of what I wanted a piece of software to do, handing them to the system, watching the system produce code that either did what I'd described or didn't, and iterating. The first application I cut my teeth on was The Trade Codex, a trading journal and education platform. From there I moved to Finaize, then SkipDay, then BidForge, each one a different domain, a different audience, a different set of problems. The first few VOSes went through revisions before they passed the verification gate, and the rejection rate dropped week over week. By the end of February I'd shipped enough of them that I'd stopped counting as a matter of course. Somewhere in there I noticed I hadn't been to a standup in three weeks, not because I'd been skipping them, but because there had stopped being standups. The methodology the book is about doesn't run them. We'd quietly stopped, and nobody had noticed, because nothing a standup had been for was not being handled by something else.

Writing a VOS is the work of a person with a good head for how a business runs and what it needs its software to do, and that was always my work. What wasn't my work, before I got here, was the step after writing it down, the step where somebody who knows how software is built takes what I wrote and produces something that runs. For the entire twenty years of my working life, that step had been someone else's job. That step is now the machine's job. And the step before it, which is my job, is the same job it always was, except now when I'm done doing it, something actually runs.

I'm not going to tell you I'm a software engineer, because I'm not. I don't write code. What I do is direct code. I can hold a conversation about how the application stores records, or why a build failed, or what went wrong in the last release, without getting lost, and I can ask the right question. I can tell when the answer I'm getting back is thin. What I can't do is type the fix. That's the line the old industry drew between my job and the engineer's job, and it's still there. What moved is the other line, the one that used to sit between my job and the code itself. The machine handles that line now, and what that's felt like, for someone who'd spent twenty years accepting that line as a ceiling, is a kind of liberation I did not see coming.

The name the methodology gives my role is *Intent Engineer*, and it turns out I'm one. I didn't plan to be one. I was hired to run operations, and I still do. I just also ship production code now, which wasn't on the job description, and which is, for reasons I'm going to spend Chapter 8 explaining, the most interesting thing that has happened to me professionally.

Short notes like this one appear at the close of nearly every chapter in this book. They're what the chapter looks like from where I'm sitting, which sometimes isn't the same place David or Glenn is sitting. Read them as you go or skip them and come back as a running commentary. They're yours. And if any of this is landing, what I'd like you to do is stay with us through the rest of the book. The methodology will keep unfolding. This note is just the first time you're hearing from me, and I'm glad you're here.

Notes

1. Ken Schwaber and Jeff Sutherland. *The Scrum Guide,* 2020 edition. scrum.org/resources/scrum-guide (accessed 2026-04-22).

2. Eliyahu M. Goldratt and Jeff Cox. *The Goal: A Process of Ongoing Improvement.* North River Press, 1984 (3rd revised edition, 2014). The Herbie parable appears in Chapter 14 of the 1984 first edition.

CHAPTER 2. YOUR FRAMEWORK IS THE BOTTLENECK NOW

In the last chapter I said the bottleneck moved. This chapter is about the machine that sits on top of the bottleneck that used to be there and is still sitting on top of it even though the thing it was built for is gone. The machine is the framework: Scrum, SAFe, their cousins, their scaled variants, their enterprise cousins, their industry-specific tailorings. They were designed for a constraint. The constraint has changed. The machine has not.

What follows is a careful, specific, and calibrated argument. I have spent twenty years watching enterprises adopt agile frameworks. I helped some of those adoptions happen. I have respect for what these frameworks were built to do, and I will pay that respect in the first section of this chapter, because a book that dismantles something it never took seriously has not earned the dismantling. Then I will name what has broken, and I will show the counting.

This is not a temperamental argument. It is a structural one.

1. What Scrum got right

Scrum's first formal articulation landed in 1995. The Agile Manifesto followed in 2001. Those two documents, and the practice that grew around them, were a correct response to a real problem. Software in 1995 was built by humans typing code into editors, and the typing was the long pole of the schedule. Coordinating a room of typists so their output lined up was the hard part of running a team. Plans were wrong the day they were written because requirements moved faster than the typing could keep up. Waterfall had already failed to hold, not because anyone was stupid, but because the central assumption inside it (plan exhaustively, then execute) did not survive contact with markets that changed every quarter.

Scrum's insight was modest and correct. Slice the work small. Inspect and adapt on a short cadence. Get the people doing the work into the same conversation as the people asking for it. Do not pretend you know what will be true in month nine; plan two weeks at a time and let the feedback from shipped increments update the plan. That insight was right. It is still right, for the constraint it was designed against.

The practices that followed (standup, sprint planning, review, retro) were the coordination moves a careful engineering culture gave itself when typing was expensive and humans were the constraint. Schwaber and Sutherland's *Scrum Guide* [1] is the canonical statement of those practices, and the framework I'll be quoting against in the sections below. The fifteen-minute circle existed because status information was otherwise trapped in people's heads and decayed fast. Sprint planning existed because a two-week commitment to a bundle of work was a reasonable bet against a horizon nobody could see past. Retro existed because the machinery of people building things together is always a little bit broken and needs deliberate attention. None of that was absurd.

I want the reader who has spent a decade running Scrum to hear this first. The framework you have been running was a good answer. It stopped being a good answer not because the people running it got worse, but because the problem it was answering stopped being the problem.

2. *What Scrum assumed*

Every methodology carries a set of implicit beliefs about what is scarce, what is expensive, and what is reliable. Scrum's are easy to list if you are willing to look at them directly.

Humans are the slow step. The thing that takes time in software is writing the code. If you can coordinate the humans better, you will ship faster. The whole ceremony stack is downstream of this assumption.

Scope is negotiable. You can't know what you'll actually need to build past the next two weeks, so you slice the work small, ship it, and let the product manager adjust. This is a correct move when the implementation cost is high. When implementation is cheap, scope negotiation becomes a different question.

Integration is expensive. Pieces of software written by different people do not line up by default. The larger the group, the worse this gets. Hence pairing, continuous integration, sprint-long integration branches, the whole mental model of managing merge risk. This one had teeth when typing dominated. Less so now.

Code is the artifact. The thing you are building, the thing you review, the thing you version, the thing you argue about, is the code. Requirements and designs are upstream ephemera. They are notes on the way to the real thing. This is the assumption that most quietly shapes everything else.

Verification is downstream. You build the thing, then you check it. QA is an organizational function that runs after development, often in a separate team, often on a separate schedule. The test suite is a safety net placed under the typing. It catches the typing's errors.

A side note on this last assumption. The regulated industries have been quietly disagreeing with Scrum on it for thirty years. IEEE Standard 1012, Independent Verification and Validation, requires the verification function to sit independent of the build team and to operate on both ends of the pipeline. The validation half checks that the requirements actually express the intent before the build begins. The verification half checks that the build satisfies the requirements after it ships. Casey and I spent five years practicing IEEE 1012 against major state systems-integration programs in Texas, and the structural shape of CID is recognizable inside that work. The regulated industries got the separation right. Mainstream Agile flattened it into a single in-team QA function and lost most of what made it work. CID restores the separation, sized for a three-person team rather than a state agency.

Now read those five assumptions again with AI generation in the picture. Humans are not the slow step anymore; specification is. Scope is not negotiable in the old way, because the implementation cost that made scope a hard thing to estimate has collapsed. Integration is a different kind of expensive, because what is merging is no longer small units of human judgment but large, rapidly-produced bodies of generated work that need a different check. Code is still what runs, but code is no longer the authored artifact; the specification is. Verification cannot be downstream of generation that operates at generation speed, because the downstream will be swamped by the output of the upstream.

Five assumptions. Four of them have inverted under the new constraint. One of them (code is what runs) is still true but answers a smaller question than it used to.

This is the substrate under the coordination machinery. The machinery itself is fine. The substrate it was built on has cracked.

SAFe role	What it coordinates in SAFe	CID/ELCID equivalent
Product Owner	Team-level backlog, sprint scope	Folded into Intent Engineer
Scrum Master	Ceremony facilitation, impediment removal	No equivalent. The ceremonies are gone.
Developer	Implementation	Folded into AI Orchestrator
Product Manager	Program-level backlog, PI objectives	Folded into Intent Engineer

System Architect	Cross-team technical alignment	Folded into AI Orchestrator at stream scope
Release Train Engineer	Coordinates PI cadence for an Agile Release Train	No equivalent. Continuous flow replaces PI cadence.
Business Owner	Approves PI objectives, accepts increments	Intent Stream sponsor in ELCID portfolio
Epic Owner	Shepherds epics through portfolio Kanban	Folded into Intent Stream owner
Solution Manager	Backlog for a Solution Train (multi-ART)	Folded into portfolio-level Intent Engineer
Solution Architect	Solution-Train-level architecture	Folded into portfolio-level AI Orchestrator
Solution Train Engineer	Coordinates Solution Train cadence	No equivalent.
System Team	Shared infrastructure and integration	Folded into platform pod; not a per-stream role
Shared Services	Horizontal capabilities (security, DBA, UX)	Verification Owner absorbs the security pieces; the rest stays as platform services
Lean Portfolio Management	Funding and governance across ARTs	Enterprise Verification Officer plus portfolio council; the governance structure of ELCID

Table T2 · SAFe roles mapped against CID/ELCID.

Six roles in CID/ELCID. Fourteen in SAFe. The reduction isn't ideological. It's structural. Roles that coordinated the old bottleneck have no work to do in the new one. Roles that absorbed what's left got redrawn around the artifact that now matters, which is the VOS.

3. The ceremony audit

Let me do the arithmetic without rhetoric. A ten-person Scrum team in a two-week sprint. I will count only the ceremonies that are in the published framework, not the extras that accrete around them.

Daily standup: fifteen minutes, ten people, ten working days in a sprint. That is ten hours of team time per sprint, and I am being generous by calling it fifteen minutes.

Sprint planning: four hours, ten people. Eight team-hours. Some shops split this into a What meeting and a How meeting; if you do, the number goes up.

Backlog refinement (sometimes called grooming, which I am not a fan of as a word): typically two hours per week, ten people. Six team-hours per sprint, counting two sessions.

Sprint review: two hours, ten people, plus stakeholders. Four team-hours, not counting the stakeholders' time, which has a cost somebody else is paying.

Retrospective: two hours, ten people. Four team-hours.

That is thirty-two hours of coordination overhead per sprint, against eighty hours of sprint capacity per person. Forty percent of the working time of the team is spent on the machinery of running the team. And I have not yet added Scrum of Scrums for any team dependent on another team, Product Owner sync, the architecture review huddle, the design critique, the dependency-planning meeting, or the two hours a week a good Scrum Master spends fielding questions the ceremonies did not cover.

If you run SAFe, add PI Planning every ten weeks: two full days, every attendee present, often flown to a single location, often in the hundred-plus person range. If you price out the loaded labor of a single PI Planning event for a mid-sized organization, you are in six figures per event, four events per year, before you have shipped anything.

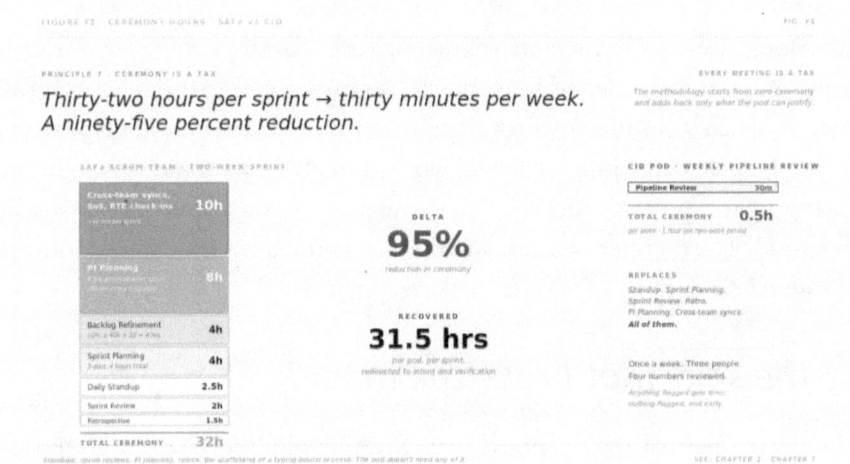

Figure F1 · Ceremony hours per sprint. *32 hours versus 30 minutes per week.*

Forty percent of a team's time was a reasonable tax when it was the price of keeping ten people aligned around the production of code that cost days-per-feature to write. The cost of the feature was the dominant number; the cost of coordination was the tax you paid to get that dominant number to land on target. When the dominant number shrinks by an order of magnitude, the tax does not shrink with it. The tax stays exactly the same size, because the ceremonies are timeboxed by calendar, not by the cost of the work they are coordinating. A fifteen-minute standup is fifteen minutes whether the team is producing a feature a day or a feature an hour.

I find this genuinely irritating. Not because the ceremonies are badly designed, but because they are being run in a context they were never calibrated for, and the people inside them can tell. A standup for a team whose implementation layer is a machine is a room full of people reporting on something that does not need reporting. The failure is not in the people. It is in the design of the room.

One paragraph of dry indignation is what I was allotted here, and I have used it. Back to the counting.

Thirty-two hours per sprint times twenty-six sprints a year is eight hundred and thirty-two hours per ten-person team per year spent on ceremony. For a fifty-person engineering organization that works out to something on the order of four thousand hours a year, or roughly two full-time-equivalent positions worth of time, spent coordinating the typing that is no longer happening. I have worked with organizations where the ceremony footprint was double that number, once you added the scaled-agile overhead. The numbers are real and they are countable. They were countable before AI showed up. They are embarrassing now.

4. The shift that broke the model

Here is the part of the chapter I have been building toward, so I will slow down and get it right.

Between roughly the middle of 2023 and the end of 2025, a quiet thing happened inside a lot of engineering teams, including the one Glenn and I were running at Alchemaize in the second half of 2025. The thing that happened is that the rate of correct code generation moved from days per feature to minutes per feature, on the kind of work that used to dominate a sprint backlog. This did not hold for every domain or every task, but it held for enough of the work, and enough of the time, that the ratio of typing cost to coordination cost inverted.

The inversion is the whole point.

Under the old ratio, typing dominated, coordination was the tax, and the frameworks were rationally calibrated to hold the tax low relative to the typing. Under the new ratio, typing is cheap, coordination is dominant, and the same frameworks have inverted their purpose. They are no longer coordinating expensive work at low overhead. They are coordinating cheap work at high overhead. The machine has the same shape. The environment has flipped around it.

A framework that runs thirty-two hours of coordination against feature work that takes minutes instead of days is not a framework that is slightly miscalibrated. It is a framework that has inverted its function. It is now more expensive than the thing it exists to coordinate. That is the definition of a framework that has become the bottleneck.

I want to state this as plainly as I can. This is the claim the chapter is organized around, and the rest of the book is the work of making it useful. **When the coordination cost of your framework exceeds the implementation cost of the work it coordinates, the framework has inverted from lift to load, and it is now the structural constraint on your throughput.** That is the pull quote. I mean it as structural observation, not indictment.

I watched this happen in real time at Alchemaize in the fall of 2025. Glenn and I had started building Ember with AI-augmented tools (Cursor, Kiro, a mix of others), still inside what I would generously describe as a reasonable Scrum-adjacent practice. We were running short standups, we were slicing work, we were reviewing at the end of sprint. By November, the standups had started to feel strange. We were reporting on features that had already shipped before the standup started. We were planning sprints for work that would complete inside the first two days of the sprint. We were coordinating an implementation layer that no longer needed coordinating, because the implementation layer was now a machine answering to whichever one of us had written the clearest specification that week.

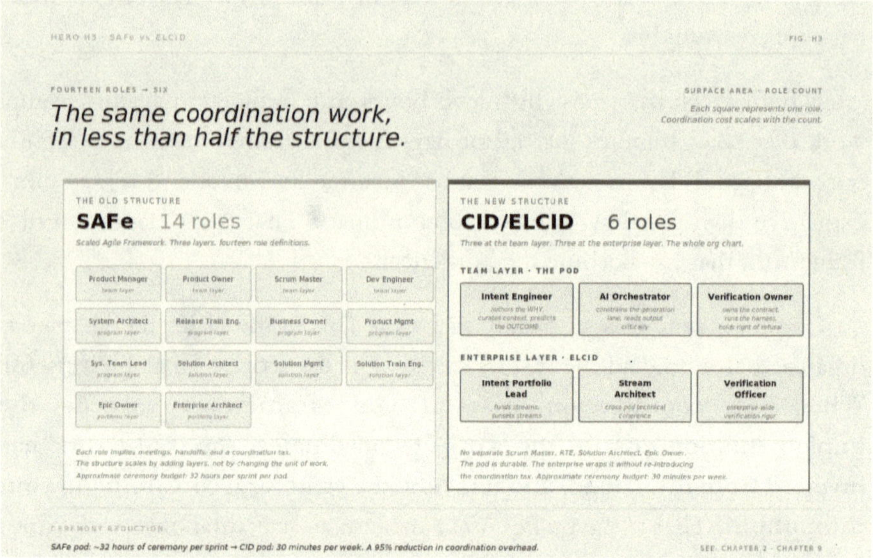

SAFe vs ELCID. *Fourteen roles reduced to six, with the surface-area geometry visible.*

The honest thing to do when you notice this is to stop running the ceremonies and start running a different shape of work. The less honest thing, which I have watched a lot of enterprise teams do in the last twelve months, is to keep running the ceremonies and add AI tooling inside them. That combination is worse than either option alone. You are paying forty percent ceremony overhead on work that now takes a tenth of the time, which means the ceremony has become sixty, seventy, eighty percent of the effective cost of delivery. You have accelerated implementation and left the coordination layer untouched, and the coordination layer is now the dominant cost of the pipeline.

The framework did not adapt. It amplified the wrong thing. That is the shape of the failure, and the shape of the failure is the shape of the opportunity.

There is a sharper way to state the geometric difference between the old frameworks and the new one. Scrum and SAFe are loops. The sprint is a fixed-duration loop on a single artifact across iterations. The Program

Increment is a longer loop at the portfolio level. PDCA is a loop. OODA is a loop. The shape of every framework that has tried to coordinate software in the last twenty-five years is the same shape: take an artifact, run it through a fixed cadence, return to the start, run it again. CID is not that shape. CID is a forward pipeline per VOS, with two reverse edges that exist for fast correction inside one trip, plus a watching layer that runs in parallel against the deployed inventory and seeds new intents at the head of the stream. The pipeline ships its artifact and is done. The watching layer keeps the system honest at a system-level cadence the pipeline does not see. That is kaizen, drawn the way Toyota actually drew it on the factory floor, not the way the methodology textbooks redrew it as a wheel. CID inherits the geometry. The book is about why that geometry, applied to software, is the framework the new constraint requires.

5. The framework as load, not lift

The design purpose of a framework is to remove coordination load so that the skilled work can happen. That is what Scrum did in 2001. That is what Waterfall tried to do in 1970. Every useful methodology in the history of software engineering has had the same job description: take the tax of coordinating a group of people off the shoulders of the people doing the work, so the people doing the work can do more of it.

The argument between capability and ritual is an old one. Nicole Forsgren, Jez Humble, and Gene Kim's *Accelerate* [3] made the empirical case years before AI generation arrived: what separates high-performing engineering organizations from the rest is a set of measurable capabilities, not the ceremonies they wrap around those capabilities. A team can run every Scrum event on the calendar and still deliver slowly; another team can run almost none of them and ship every day. The capability is what carries the work, and the industry has spent twenty years mistaking the ritual for it. Once you have that distinction in hand, the inversion below is easier to see.

When the tax gets heavier than the work, the framework has inverted its job. It is not removing load; it is adding load. It is not lift; it is ballast.

There is a mechanical test you can run on your own organization. Count the hours of ceremony your team spends per unit of shippable output. Count the hours of generation (including verification) your team spends per unit of shippable output. Compare the two numbers. If the ceremony number is larger than the generation number, your framework is load, not lift. If the ceremony number is more than double the generation number, your framework is the dominant cost of shipping software at your company, and the AI tools your developers are using are carrying the generation cost alone while the ceremony stack eats the budget.

I will give you a concrete example from our own ledger. In February of 2026, Casey shipped his fourteenth VOS against Finaize. The specification took him about an hour to write. The generation-and-verification phase took about forty minutes of elapsed time, of which maybe ten minutes was his hands on the keyboard. Total of his working time on that VOS: something close to seventy minutes. The week that VOS shipped, our entire ceremony footprint was thirty minutes. The ceremony-to-delivery ratio on that feature was under half. If we had been running Scrum the way we had been running it in October of 2025, the ceremony footprint for a team of three across that week would have been somewhere in the range of four or five hours, against seventy minutes of work. The ceremony tax would have been four to five times the work it coordinated.

That is the inversion, rendered in dollars and minutes.

The trap here is that the inversion is invisible to teams that do not count their ceremony hours honestly. It is very easy to believe that the ceremonies are cheap because they are already on the calendar and nobody thinks about them as cost. Calendar cost is real cost. The thirty-two hours per sprint is on somebody's P&L whether you call it a meeting or a line item. You are paying for it either way.

6. What's wrong with the scaling frameworks

The scaling frameworks are plural. SAFe is the one most enterprises this book is trying to reach have adopted, and it is the one I will spend the most time on. LeSS, Spotify, and Nexus are in the same conversation, and the structural argument about the new constraint applies to each of them. I will come to them after the SAFe-specific case.

The reader who is running Scrum at a ten-person team level can probably do their own arithmetic from the earlier sections of this chapter. The reader who is running SAFe at an eight-hundred-person organization has a bigger problem, and I want to be specific about what it is.

I should also say up front, before any of the criticism, that Casey and I are not coming at the scaling frameworks from the outside. Both of us are Certified SAFe Program Consultants, the highest practitioner-grade SAFe credential short of becoming a trainer, the SPCT level. We taught all of the SAFe courses. Across our consulting careers we trained and certified hundreds of Scrum Masters, Product Owners, and SAFe practitioners, including inside Fortune 100 enterprise transformations whose adoption programs we led. We did not adopt SAFe because we were told to. We taught it because we believed in it, and we believed in it because at the time we were teaching it, it was the answer the question was asking for. The criticism here is not the criticism of an outsider. It is the criticism of two people who spent years inside the rooms SAFe most needed to land in, and who now believe the rooms have changed.

With that on the record, the SAFe-specific case.

SAFe, per Scaled Agile's own reference pages at `scaledagile.com` [2], defines fourteen roles across its levels: Release Train Engineer, Solution Train Engineer, Product Management, Product Owner, Business Owner, System Architect, Enterprise Architect, Scrum Master, Lean Portfolio Manager, Epic Owner, Solution Architect, System Team, Shared Services, Customer. I may have the count slightly off depending on which version of the framework you are reading; the number has moved between 12 and 16 depending on the year. The number is not what matters. What matters is the function each role performs, and whether that function is still necessary under the new constraint.

Some of the roles coordinate across teams whose outputs need to integrate. Under the new constraint, integration is a different problem, handled by the verification gate rather than by cross-team coordination meetings. Some of the roles coordinate the cadence of a ten-week Program Increment. Under the new constraint, the PI cadence is a calendar artifact with no remaining engineering justification. Some of the roles manage the ceremonial scaffolding of a multi-hundred-person organization moving on a shared cadence. Under the new constraint, the cadence that matters is continuous, and the ceremonial scaffolding does not have a constraint to answer to.

The PI Planning event itself is the most specific case I can make. A two-day event, every ten weeks, with a hundred-plus people in a room (or on a video call approximating a room), is a ritual calibrated to produce commitment and alignment across teams whose work needs to integrate over the following ten weeks. The ritual makes sense if you are operating on days-per-feature and your ten-week horizon contains roughly forty to fifty features per team. The ritual does not make sense when the implementation layer can produce forty to fifty features in a week, because the ten-week horizon has already moved three times before the PI Planning event ends.

Glenn has a short observation I want to include here, because the PI Planning point is one he and I have argued about, and he is right in a way I was not at first.

> **Glenn, on PI Planning as ritual.** I have sat through three of these events. The function is not planning. The function is alignment theater. The output is not a plan that survives contact with the next ten weeks; the output is a shared feeling that everyone saw the same pictures on the same walls on the same two days. That feeling is worth something, at the old cadence. At the new cadence it's worth nothing, because the thing that used to create alignment (a slow-moving ten-week plan everyone agreed to) no longer exists. You cannot align an organization around a plan whose horizon is shorter than the event that creates it.

The fourteen roles of SAFe map onto the new structure in three ways. Some fold into the three CID pod roles. Some fold into the three ELCID enterprise roles. Some do not fold anywhere, because the function they performed has no equivalent under the new constraint. Table T2 earlier in this chapter shows the role-by-role mapping in detail; what I want to do here is name the structural shape behind it, because the table is correct, but the table on its own does not explain why the reduction works.

The reduction works because each of the fourteen SAFe roles existed to coordinate one of three things, and only one of those three is still expensive enough to need its own seat.

The first thing the SAFe roles coordinated was *the typing*, which is the work that used to be expensive and is now not. Developers, Product Owners, Scrum Masters, Release Train Engineers, Solution Train Engineers, the System Team, parts of Shared Services. All of these existed because writing code in a coordinated way across humans took a kind of choreography. Under the new constraint, the choreography has dissolved. A pod of three runs the typing-equivalent work continuously, with a verification gate doing the integration that the train cadence used to do socially. The roles that coordinated the typing either fold into the pod or do not fold anywhere.

The second thing the SAFe roles coordinated was *the specification*, which is now the dominant cost rather than the residual one. Product Manager, Business Owner, Epic Owner, Solution Manager, parts of Product Owner, parts of System Architect and Solution Architect. All of these existed because somebody had to be responsible for what the work was supposed to do. That work has not gone away. It has gotten more important. It has also gotten more concentrated, because a specification that was passed across a four-layer SAFe hierarchy now runs through one Intent Engineer and one Intent Portfolio Lead. The roles that coordinated the specification fold into those two seats, plus the Stream Architect for the cross-stream architectural choices the specification rests on.

The third thing the SAFe roles coordinated was *the trust*, which is the part that nobody names directly because nobody quite admits a methodology runs on it. Lean Portfolio Management, the audit function that ran inside Shared Services, the cross-team review meetings the framework treated as integration ceremonies but which were actually trust-rebuilding ceremonies. The trust function is what the Verification Officer absorbs at the enterprise layer. The standard the Verification Officer sets is what makes a sign-off a real signal rather than a rubber stamp. The role exists because the previous methodology distributed the trust function so widely that no single seat carried the authority to refuse a release. The new methodology consolidates that authority into one seat, outside the delivery chain, with the standard explicit and the harness mechanical.

Fourteen roles down to six, with one new seat (the AI Orchestrator) added because the function did not exist before. The reduction is not a cost-cutting exercise. It is a description of what the new constraint actually requires to run.

The same diagnosis lands on the other scaling frameworks. The proportions differ. The shape doesn't.

LeSS, Bas Vodde and Craig Larman's Large-Scale Scrum [4], was designed as an explicit reaction to SAFe's role-and-ceremony bloat. LeSS keeps a single Product Owner and a single Product Backlog across multiple Scrum teams, and the practitioners I worked with mostly preferred it on the grounds of parsimony. The structural argument still lands. LeSS is a coordination layer for multiple teams of human typists working a single backlog. The backlog is the work that has gotten cheap. The coordination is the work that hasn't moved.

The Spotify model, which copied around the industry from Henrik Kniberg's 2012 internal-engineering white paper [5], never actually worked at Spotify the way it was packaged for everyone else, a fact the company itself eventually acknowledged in print. Squads, tribes, chapters, guilds. The vocabulary is appealing. The structural problem is the same. Spotify's model was designed to coordinate cross-functional humans whose typing was the bottleneck. The team unit was a typing unit. The guild was a typing-knowledge-sharing structure. None of it was built for what comes after the typing bottleneck dissolves.

Nexus, the framework Ken Schwaber put together as the official Scrum.org answer to scaling questions [6], is the most parsimonious of the four. Three to nine teams, one Nexus Integration Team, one shared Sprint, one Sprint Goal. It carries the lightest ceremony footprint of the scaling frameworks, and it is also the one most exposed when the bottleneck moves. Nexus Sprint Planning, Nexus Daily Scrum, Nexus Sprint Review, Nexus Sprint Retrospective. Every one of those meetings is an answer to a coordination problem that diminishes when the implementation step diminishes.

A reader will notice I did not save my hardest argument for the framework I had the most credentials in. That's on purpose. The criticism is structural, and structural criticism does not survive selective application.

The criticism in this chapter is not that the scaling frameworks were badly designed. Each was designed well for a problem that has changed. The criticism is that continuing to run any of them against a problem that no longer exists is not neutral. It is a structural impediment to any organization trying to absorb the new implementation capability at the speed the capability actually operates.

7. The diagnostic question

If you are the reader who took the chapter at face value, you now have a question to answer, and I want to hand it to you cleanly so you can carry it into the rest of the book.

The question is the one every CTO, every VP of Engineering, every Head of Platform I have talked to in the last six months has been walking around with and not quite naming. If your team is running Scrum, SAFe, LeSS, Nexus, or any flavor of the Spotify model today, and your engineers are using AI coding tools (they are; the survey data is decisive), how many hours per sprint are you spending coordinating work that no longer needs coordinating at that cadence, in that venue, with that many people? Count the standup. Count the planning. Count the refinement. Count the review. Count the retrospective. Count the cross-team syncs. Count the PI Planning if you run one.

Then count the hours per sprint your team spends authoring specifications, curating context for generation, and running verification against the output. Compare the two numbers.

If the first number is larger than the second, your framework is not supporting your team. Your team is supporting your framework.

That is the diagnostic. It does not take a consulting engagement to run. It takes an honest look at your team's calendar for the last sprint and a willingness to accept what the arithmetic says.

The next chapter is the evidence chapter. In it, we will show what we built at Alchemaize in the hundred days between January 18 and April 20 of 2026, with a framework designed for the new constraint instead of the old one. Casey shipped across several production applications during that window, starting from zero prior coding experience. The framework made that possible. The frameworks named in this chapter would have made it impossible. The methodology in the chapters after evidence is the one we actually ran.

The framework is the bottleneck. It is also the thing you can change.

Field Note from the Test Pilot

> *Casey Robinson, Intent Engineer*
>
> **The morning I realized I didn't need a standup anymore.**
>
> The second Monday in February, I opened my laptop, looked at the clock, and it was 9:30 AM. For most of my career, 9:30 was standup time. Every team I'd been on for the last several years

had a standing meeting right around that time, fifteen minutes long, round-robin, what did you do yesterday, what are you doing today, any blockers. I'd done that meeting hundreds of times. It was muscle memory. At 9:30 my brain expected to be in a meeting.

But I wasn't in a meeting. I was at my desk in Claremore, working on a VOS for The Trade Codex. A small one, a page that let a user see their last ten entries in a table, sortable by a couple of columns. I had written the specification the Friday before, David had given me some feedback, Glenn had flagged something in the acceptance contract that needed to be tighter, and the rewrite had taken maybe twenty minutes. On this Monday I was on the third pass of generation against that VOS. The first pass had missed the sort behavior entirely, the second pass had added the sort but broken the timestamp column, and the third pass was the one that went green. It shipped at about ten-thirty, and I had a cup of coffee and a piece of working software before the old standup time would have been over.

I want to be careful about how I describe what that felt like, because it wasn't defiant and it wasn't a gesture. I just noticed that the rhythm I'd carried from my previous career, the rhythm where 9:30 means meeting, didn't apply anymore. I had nothing to report that the standup format was built to surface, and I had a piece of work in front of me that was going to take the full time the standup would have taken, and the piece of work was going to ship something.

The next morning, same thing. 9:30 came and went. I was writing a specification for the next VOS. No information I needed to share that couldn't have been shared better by the state of the repository.

By Thursday, I realized I hadn't thought about a standup in four days. And I started thinking about why. I think it's because the standup, in my previous career, was doing less work than anyone inside it had thought it was doing. The information it surfaced was mostly information the dashboard already had. The alignment it created was mostly alignment that had already been created by the previous week's work. And the main thing it produced was the feeling that alignment was being maintained, which was real but which wasn't attached to the meeting itself.

I want to be fair. There were people on teams I'd been part of who needed the standup. A new engineer two weeks into the team got real information from hearing what everyone else was doing. A PM with four squads to cover used the round-robin to notice which squad was quiet and why. For those people, the standup was doing real work, and I'm not the standup's enemy. But for someone in the role I'm in now, writing VOSes and running them through the methodology, the standup doesn't have a job to do.

One more thing, which is the thing I did not know how to say at the time and can almost say now. The standup was a ceremony for a delivery model I had already moved past. The delivery model I was in the middle of living, specification and generation and verification and done, did not have a ceremony for Monday morning. It had a queue. The queue was on my laptop. I looked at the queue every morning, and I did not need fifteen minutes of other people to tell me what was in it.

That realization was the end of my last Scrum practice, and it's funny looking back, because I didn't announce it. I didn't hold a retro on it. I just noticed one morning that the habit had stopped, and the work hadn't missed it. The VOS queue is on my laptop. The work is on my laptop. The review process is on my laptop. If you want to know what I'm doing, you can look at the queue. I can show you. It'll take less time than the standup, and what you'll see is the actual work, which has been the thing I wanted people to see in those meetings the whole time.

Notes

1. Ken Schwaber and Jeff Sutherland. *The Scrum Guide,* 2020 edition. scrum.org/resources/scrum-guide (accessed 2026-04-22).
2. Scaled Agile, Inc. *SAFe 6.0 for Lean Enterprises.* Scaled Agile, 2023. scaledagile.com/safe-big-picture (accessed 2026-04-22).
3. Nicole Forsgren, Jez Humble, and Gene Kim. *Accelerate: The Science of Lean Software and DevOps.* IT Revolution, 2018.
4. Craig Larman and Bas Vodde. *Large-Scale Scrum: More with LeSS.* Addison-Wesley Professional, 2016. less.works (accessed 2026-04-29).
5. Henrik Kniberg and Anders Ivarsson. *Scaling Agile @ Spotify with Tribes, Squads, Chapters & Guilds.* Spotify white paper, October 2012. The model spread industry-wide on the back of this paper. Spotify itself acknowledged in subsequent talks that the framework was a snapshot of an evolving practice rather than a stable target state.
6. Ken Schwaber. *The Nexus Guide,* 2021 edition. scrum.org/resources/nexus-guide (accessed 2026-04-29). The Nexus framework is the official Scrum.org answer to scaling questions and is structurally the lightest of the major scaling frameworks.

CHAPTER 3: THE EVIDENCE

In the hundred days between January 18 and April 20, 2026, three people working out of home offices in three states shipped thirty-five production applications. No shared office. No outside funding. One pod, operating at roughly two full-time-equivalents across three seats, running the methodology the rest of this book describes.

Those were the second hundred days. The first hundred days, which preceded them, were not the case study. They were what produced the case study, by failing first.

That's the number. It's the number this chapter exists to defend, and the one a hostile reader will want to audit before they give the methodology another page of attention. I've been on the receiving end of enough enterprise pitches to know how a number like that usually gets handled in a book. Big headline, vague footnote, four chapters of claim-laundering on top of it, and by the end the reader has stopped trusting the author. I don't want to do that to you. So this chapter lays the ledger down flat.

What I'm going to give you is what we built, what we counted, what we killed along the way, and what we refused to claim because we couldn't defend it yet. When you're done with the chapter, you should be able to do one of two things. You should be able to believe the number because you've seen how it was assembled. Or you should be able to disagree with a specific line item and know exactly where to push. Either of those is a good outcome. The bad outcome is the one where I convinced you of a feeling and you walked away without the parts.

This is the evidence chapter. It's the heaviest one in the book. Read it with a pen.

---

1. What the chapter will and will not claim

I want to stake out the posture of the chapter before the data starts, because the posture is the thing that makes the data legible.

This chapter will claim that a three-person team with one non-engineer shipped thirty-five production applications across fourteen weeks using a named methodology, and that the output is auditable down to the repository level. It will claim that the throughput multiplier we measured against an industry baseline sits between thirty and fifty times the traditional rate, for this specific team, on this specific body of work, in this specific period. It will claim that the methodology's internal quality discipline (verification upstream of generation, acceptance contracts before code, context curation as a first-class practice) is what kept the speed from collapsing into debt.

The chapter will not claim that AI replaces engineers. Two of the people on the pod are engineers or adjacent, and the non-engineer (Casey) was operating inside a methodology whose verification seat was held by a career architect with two US patents. The claim the number rests on is not "anyone can do this." The claim is "this team did this, under these conditions, using this method, and here is the audit trail."

The chapter will not claim that the methodology is easy. Casey's first days were boring in the way learning any new discipline is boring. A four-hour first specification, Glenn rejecting most of its acceptance contracts the next morning, the slow climb to forty-five-minute authoring cycles with a defensible reject rate. Nothing in this chapter is going to tell you it feels like magic, because it didn't.

The chapter will publish the specific Alchemaize bug rate, but with the disclosures that go with publishing it. The industry benchmark we measure against is Steve McConnell's *Code Complete*, 2nd ed. [1], which establishes the canonical range at fifteen to fifty defects per thousand lines of delivered code; § 6 walks through that anchor. Our measured rate across the hundred-day sprint sits under three defects per thousand lines, which is below the lower end of McConnell's range and roughly five times better than the fifteen-per-KLOC figure most often cited as the working industry midpoint. The number is real. The disclosures are also real: the rate is measured across a specific sample of our production repositories, over a specific observation window (the hundred days the rest of this chapter covers), against a specific definition of what counts as a defect (any production incident triaged as a code defect, plus any post-deploy bug found in the first thirty days against the shipping VOS). Glenn is writing the full measurement methodology up as a companion piece, scheduled for release after this book ships, so a reader who wants to audit the number can. Until that piece lands, the right way to read our number is as an internally measured rate that is consistent with our McConnell-anchored claim, not as a peer-reviewed benchmark. The next edition of this book will fold the companion piece's findings back into this chapter.

What's left, after all the things the chapter won't claim, is a specific kind of evidence. It's evidence about what a small, well-configured pod can produce under a particular methodology in a particular window, measured against a particular traditional baseline, with disclosures attached. It isn't evidence about the future of software. It's evidence about one team's hundred days. The inferences to the future, and to your organization, come in the later chapters, carried by arguments the number only partly underwrites.

I want you to have the number and the disclosures before any of those arguments arrive.

---

2. The 100 days

The calendar first, because every other fact in this chapter sits on top of it.

The first 100 days (October 1, 2025 to January 9, 2026). Alchemaize was incorporated in August 2025 to build Ember, an AI-augmented reading application. Casey joined as COO on October 1. By October 20 we had a four-person development team in place: three developers and an AWS solutions architect, all part-time, all comped on options and deferred contingent compensation rather than salary, all holding full-time jobs at FAANG companies concurrently. We built Ember the old way: standups, sprints, retros, sprint reviews, planning. AI tooling (Cursor, Kiro) was layered on top of the traditional methodology, and the two ways of working spent most of the fall in tension over how work was characterized, executed, and released. The web-app MVP shipped on December 1. User-acquisition targets did not follow. The product worked on Project Gutenberg titles, and the target audience wanted commercial fiction, and DRM made the commercial-title catalog inaccessible without a Kindle integration the company could not yet open a conversation about. By the end of December the development team was wound down. The first 100 days produced an MVP and a clear answer: the way Alchemaize was working was not going to work.

January 18, 2026. Day one of the second 100 days. The line in the sand. By early January, David, Glenn, and Casey had concluded that Alchemaize had to fundamentally change how it worked or it did not have a future. The methodology had a name now: Continuous Intent Delivery. The pod was the three founders. No development team. No architect. Three people, three states. We were going to operate the pattern (author intent with enough precision that a machine can act on it, delegate implementation to a competent generation layer, verify upstream and continuously, repeat) across every project from this day forward and see what came out.

January 28, 2026, 4:00 AM Central. Day eleven of the second 100 days. Amazon issued a company-wide reduction in force. Sixteen thousand Amazonians received notices that their positions had been eliminated, with a 90-day notification period running from January 28 through April 28, 2026. David's entire team of 14 was hit, including David himself and Casey, who led the early-migrations and EBA sub-team within David's broader org. By 6:05 AM David had texted Casey: *Hey, I think our whole team got hit including me.* Casey woke at 8:00 to a wall of messages from his own sub-team and David's. His reply at 8:18: *Yep...* By 8:20 they were on a Zoom call together, working out how to support the rest of the team through the morning. The week ended with Alchemaize's two part-time founders converted into two full-time founders, ten days into a methodology that had not yet shipped a single Casey-authored VOS.

January 29, 2026. Day twelve. Casey's first Verifiable Outcome Slice. A three-day window, January 29 through 31, against thetradecodex.com, a small vertical tool we needed a working version of fast. The first VOS took him four hours to author, and Glenn rejected three of its four acceptance contracts the next morning. That morning was the first time any of us had watched the methodology do what it was supposed to do in the hands of someone who wasn't writing the code. The rejection rate said the verification seat was working. The fact that Casey shipped a correct VOS on his third attempt said the authoring seat was working. By the end of that weekend the specifications were already getting shorter and the reject rate was dropping.

February 1, 2026. Casey moved off TradeCodex and onto Finaize, a modernized Auto F&I web app with a mobile app companion. Finaize showed the breadth of what the methodology could handle: a completely different domain, a different user base, a different set of business rules. The learning curve that started on TradeCodex continued on Finaize, and by mid-

February Casey was authoring VOSes in under an hour with a first-pass acceptance rate most engineers would be satisfied with.

Through February. The pod ran three to five active projects concurrently. I held the AI Orchestrator seat most of the time. Glenn held the Verification Owner seat. Casey was the primary Intent Engineer, but the roles blurred. On any given day one of us might be authoring a VOS for another of our projects, because the roles are functions, not headcount. We were distributed across Brentwood, Claremore, and College Station. We used Slack for asynchronous work and scheduled video calls when a decision needed all three voices in the room at once. The room was always virtual. There has never been a shared office, and there was never going to be one.

Through March. Projects decommissioned: CrocTrade (a crypto trading dashboard we killed because the hypothesis didn't hold in user data), AegisTrader (an options platform whose market assumptions we invalidated during customer conversations), and YouTubeTranslator (a translation utility we shipped, evaluated, and retired when a better approach emerged). Each was a VOS-complete application with tests and infrastructure. The decommissioning is part of the evidence, not a subtraction from it. A methodology that lets you kill three production applications inside the window it took to ship them is a methodology where the cost of being wrong has collapsed. That collapse is itself part of what the number measures.

March 1, 2026. Alchemaize launched as an AWS partner. By this point the methodology had a name: Continuous Intent Delivery. CID at the pod layer, ELCID at the enterprise layer. Glenn and I named it the second week of February; Casey made us promise we wouldn't put the name in a deck until we'd shipped another twenty VOSes against it. We honored the promise.

April 1, 2026. CATALYST released as one of the commercial offerings in the AWS partner launch. CATALYST is the productized form of the methodology, the part a customer buys when they want the pod structure and the verification harness delivered as a service.

April 20, 2026. Day one hundred of the second 100 days. By this point the pod had shipped thirty-five production applications, roughly one million lines of production source code (counted under the rules in § 5), 429 test files, and had closed roughly two thousand commits across the portfolio. The traditional-effort estimate, using standard industry productivity baselines for senior full-stack engineers, is about 9,150 developer-days, which is roughly thirty-one developer-years. A six-person traditional engineering team working for five years would have matched the output, at a cost profile Alchemaize did not and could not have afforded.

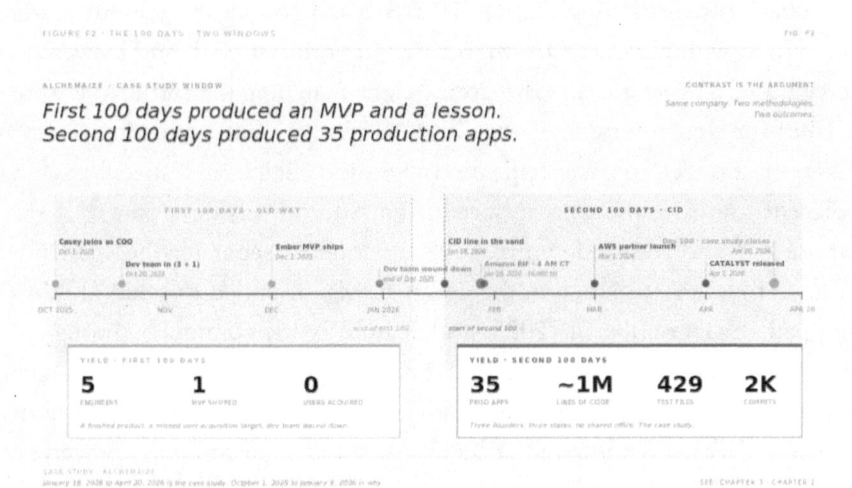

Figure F2 · 100 days, 59 shipped VOSes across the thirty-five-application portfolio. *Each tick on the timeline is a shipped Verifiable Outcome Slice; multiple VOSes typically ship into the same application, which is why the VOS count exceeds the application count.*

The hundred days is the window. The ledger in § 3 is what filled it.

---

3. The ledger

The thirty-five projects, grouped by category. Every entry is a working system that either went to production or was deployed to production and then decommissioned during the window. Nothing in this list is a prototype we never ran, a demo we never showed, or a spike we never shipped. The audit trail is the git histories of the repositories. For the evidence chapter of the book's published form we'll include an appendix with repository references; for the manuscript draft, the category-level grouping and project names below are the record.

Web platforms (eleven projects, including two decommissioned). Yeon CRM is a modern customer-relationship platform built around verticals we'd been asked to serve. STR.fish is a short-term-rental analytics and operations product. FinAIze is an AI-first cloud-native auto finance and insurance platform. AegisTrader is a Schwab-connected options trading dashboard with per-user OAuth, real-time quotes via SSE, and live account integration. Tanaiger is an AI-driven freight-matching platform for logistics, a Uber-for-freight model. Trade Codex is the AI trade journaling platform Casey cut his teeth on, a specification-ingestion engine for a specific trading-adjacent use case; the web surface at thetradecodex.com is part of a stack whose iOS and Android companions are counted separately below. Ember is the AI reading companion the company was founded to build; the MVP shipped on December 1, 2025, and feature work continued through the sprint. VisibleWealth is a financial-overview product for individuals tracking accounts, budgets, and net worth. BidForge is an AI-powered RFP response platform that helps companies decompose incoming proposals and generate structured responses, pulling in organizational strengths and identifying gaps. Plus CrocTrade (decommissioned) and YouTubeTranslator (decommissioned). Trade Codex carried the bulk of Casey's Intent Engineer work in the early window. Web platforms are the applications that carry the most VOS weight because they require the deepest specification work on the Intent Engineer side and the tightest acceptance contracts on the Verification Owner side.

Mobile applications (ten projects). Drawer is an AI-organized life organizer for iOS. NoshMode is an AI meal planning app for iOS and Android. Trade Codex mobile is the companion surface to the trading platform on iOS and Android. Ember mobile is the companion to the reading platform on iOS. VisibleWealth mobile is the iOS companion to the financial-overview platform. Renew is a daily Bible-action product for iOS and Android, turning the KJV into personalized daily challenges with a transformation journal. FlipMode is a focus product that turns iPhones into flip phones at the OS level via Screen Time. SkipDay is a no-pressure daily workout coach with pain-aware exercise selection and Apple Health integration. BoxLens is an AI inventory tool with NFC tag scanning, in TestFlight at the close of the window. Razz is a voice-driven AI roast app, in beta. Each is a React Native, Expo SDK 55, or native iOS build; each was submitted to TestFlight, the App Store, and/or Google Play; each runs against a production backend the pod also built during the sprint. The mobile category is the one that most obviously stresses the methodology's breadth claim: a traditional mobile team at this output level would have been a half-dozen engineers per app, dedicated, for a quarter. Our pod built ten mobile apps alongside everything else.

Native iOS and Mac (two projects). ThreadLens is a local-first iMessage analytics product for macOS with an iOS companion; the Mac extracts and analyzes chat.db on-device while the iOS app reads the derived artifacts. Naeum is an iOS plus watchOS holistic wellness companion tracking six dimensions of personal practice (Move, Nourish, Breathe, Focus, Connect, Rest) through an on-device Daily Rhythm Engine. Swift and SwiftUI throughout, with watchOS extensions where the surface demanded them. The native-iOS category is the one where the productivity multiplier runs lowest, for reasons the § 4 table covers: Xcode toolchain friction, device-specific debugging, and the parts of Apple's platform where AI-generated code still needs an operator in the loop for longer stretches than anywhere else in the portfolio.

Games (five projects). My Albo is a Korean-food virtual pet for iOS and Android: hatch an egg, evolve through fourteen characters across six life stages based on care quality. Radient is a retro vector arcade tube shooter on iOS and Android, an 80s-style game with sixteen web shapes, seven enemy types, super zapper, and procedural audio. Helia Sort is an iOS color and shape sort puzzle game in beta. Runeshell is a browser-based visual MUD for the modern web. Voidtrader is a vector-style space trading and combat sim served from the browser, a love letter to Elite (1984) with a procedural galaxy and seventeen-commodity trading. The games stress-tested the methodology against work types (real-time game loops, sprite rendering, browser performance, native iOS game frameworks) that the industry assumes don't respond well to AI-augmented development. The methodology held.

Tools and frameworks (two projects). CATALYST itself, the delivery framework that became our April 2026 commercial offering. And AIzoth, the internal VOS-tracking and measurement tool the pod used to count what was shipped against which application; the numbers in this chapter come out of it. Tooling is a category where the pod's work is partly recursive: the methodology building the methodology's tooling.

Infrastructure (one project). The Alchemaize landing zone, authored in AWS CDK. Not a large LOC count, but the blast radius of a mistake in this category is asymmetric, which kept the methodology honest about verification. The landing zone underpins every other application in the portfolio; a bug in an app hurts an app, a bug in the landing zone hurts the company.

Marketing and content sites (four projects). alchemaize.ai is the company site. davidkimio.com is David's author site, the companion to *Bullets Don't Fly* and *The Real Money Guide*. therealmoneyguide.com is the companion site for *The Real Money Guide* with more than thirty interactive financial tools, refactored and hardened during the sprint. Plus a product launch landing page tied to the CATALYST commercial release. Content sites are high-multiplier projects because the templating patterns are highly repetitive; they are the category a skeptical reader is going to discount first, and the counting rules in § 5 make that discount possible if the reader wants to apply it.

Thirty-five entries. Two decommissioned. Thirty-three live at the end of the window.

A word on the decommissions, because the temptation in an evidence chapter is to hide them. Every methodology book you've read has a ledger with only survivors on it. Ours has survivors and two specific failures, named. CrocTrade failed a hypothesis about retail crypto trader behavior we'd assumed would hold; user data disagreed, and we killed it in week eight. YouTubeTranslator shipped, ran for six weeks, and was retired when a simpler integration path showed up that didn't need a dedicated product around it. Neither is a story of the methodology failing. Both are stories of the methodology letting us find out we were wrong quickly enough that being wrong was cheap. That's the point of keeping them in the ledger.

The honest number is thirty-five shipped, two decommissioned, thirty-three live. If you want to discount the thirty-five to thirty-three, you can, and the argument of the rest of this book survives. If you want to count the two decommissions as methodology wins rather than subtractions, I think you're right, but I'd rather you got there on your own than have me argue into it.

---

4. The productivity multiplier, labeled three ways

The number most likely to be misquoted from this book is the multiplier. I'm going to stake out three distinct multipliers, three distinct scopes, and three distinct rhetorical jobs, because the three get conflated in every conversation about AI productivity I've been in for the last year, and I want the book to be the place that stops doing that.

Multiplier one: 30–50×. Alchemaize-specific. This is what the Alchemaize pod measured on the hundred-day sprint, across the thirty-five-project portfolio, against a traditional-baseline estimate of senior full-stack engineer productivity. It's a blended average across work types. It is not a claim about the industry. It is not a claim about every team. It is a claim about one pod on one window of work, measured under the counting rules in § 5. When you see 30–50× in this book, it means "what we did." Nothing else.

Multiplier two: 10–40×. Methodology-level. This is the broader claim. AI-native development, done with the discipline the methodology describes, produces output at roughly one to two orders of magnitude above the traditional baseline. The range sits lower than the Alchemaize-specific number because it accounts for teams that are newer to the methodology, teams working in harder-to-automate stacks, and the general variance that shows up when you sample broadly. When you see 10–40× in this book, it means "what the methodology does, in general." It does not mean "what Alchemaize did."

Multiplier three: 5×. Conservative scaling. This is the multiplier the book uses in Chapter 13 when it runs enterprise CIO projections. It's deliberately conservative. Five times the traditional baseline, sustained, over a calendar year, accounting for coordination overhead, onboarding, context switches, and all the ways enterprise reality eats into pod-level numbers. If you're a CIO reading this book and your instinct is "I don't believe thirty times," I don't need you to. The scaling math in Chapter 13 works at five. When you see 5× in this book, it means "the floor we'll defend, at scale, against the most skeptical reader in the room."

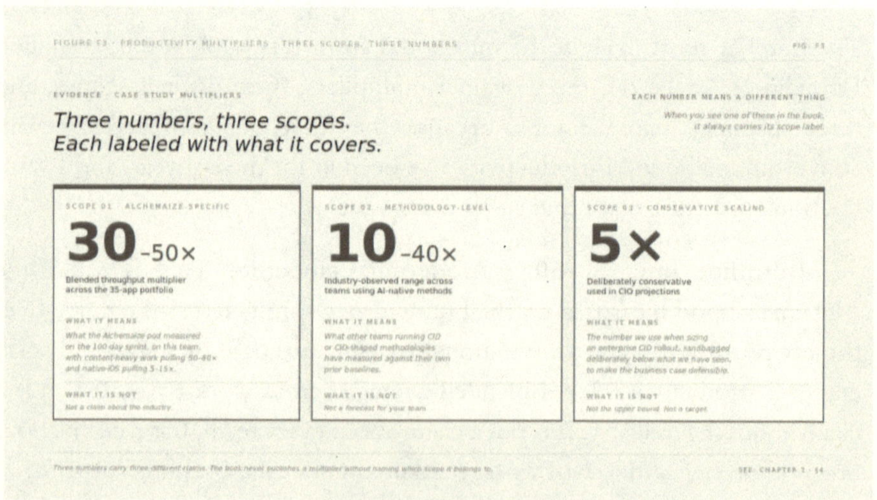

Figure F3 · Productivity multipliers. Three scopes

Three multipliers, three scopes. Every time one of them appears in this manuscript, it carries a scope label. 30–50× is always labeled Alchemaize-specific. 10–40× is always labeled methodology-level. 5× is always labeled conservative scaling. We're disciplined about this. If we drift, catch us.

Inside the Alchemaize-specific 30–50× blended number, the work-type variance matters. The multiplier is not uniform across the portfolio. Here's the breakdown by work type, as measured internally.

Work type	Multiplier range	What drives it
Content sites and templating	50–80×	Highly repetitive patterns, component composition, near-perfectly automatable
CRUD web applications	30–50×	Standard patterns for forms, tables, API routes, database queries
Mobile app UI	25–40×	React Native component trees, navigation, state management follow predictable patterns
Backend APIs and business logic	20–35×	Route handlers, middleware, auth flows, data transformations
Infrastructure as Code	10–20×	CDK constructs are well-documented, but blast radius forces human judgment
Native iOS / platform debugging	5–15×	Xcode toolchain issues and device-specific bugs still require iterative human work

The blend is 30–50× because the portfolio has that mix. If you took the same pod and pointed it exclusively at native iOS work, the multiplier would drop to the bottom of the range. If you pointed it exclusively at content sites, the multiplier would sit at the top. The methodology scales with the repeatability of the work, which is unsurprising. What's surprising is that even at the bottom of the range, the multiplier is still five times what a senior engineer would produce unassisted. There is no work type in our portfolio where the methodology did not outperform the traditional baseline.

A word from Glenn on the measurement, which I've been promising for three sections now.

> Measurement rule: we count what a human could have authored unassisted, divide by what the pod actually shipped under the methodology, and take the ratio. We don't count what the AI "generated" as gross output, because gross generation includes boilerplate a human would have scaffolded from a template in the old world. We count shipped, tested, verified production code, same as we'd count it if a human had typed it. The ratio is honest because the numerator and denominator are the same kind of thing. We're not comparing AI-generated kilobytes to human-written kilobytes. We're comparing what shipped under one method to what would have shipped under another, with the same quality bar on both sides. The bar is set by the acceptance contract, not by the method. That's what makes the multiplier a real multiplier and not a vanity metric.

Back to me. The multipliers are the single most-likely thing in this book to be pulled out of context, turned into a slide, and used in someone's all-hands at a company we've never met. If it happens, please put the scope label next to the number. 30–50× Alchemaize-specific. 10–40× methodology-level. 5× conservative scaling. The scope is the number's truth.

A worked example, because the abstraction needs a concrete floor.

Take TradeCodex. The application that carried the most VOS volume for the Intent Engineer seat during the sprint. TradeCodex has, at the end of the window, roughly 42,000 lines of application source code under the counting rules in § 5, across 211 functions and components, 58 test files, and about 340 commits. If you asked a traditional estimation model how long a senior full-stack engineer would take to produce that output, at the same quality bar (authored tests, production-grade data handling, real market data integrations, customer-facing UI), the answer sits between 120 and 180 engineer-days. Call it 150 as a midpoint. That's roughly 30 engineer-weeks. A single senior engineer, dedicated to TradeCodex alone, would need about seven and a half months to produce the current application from scratch.

The pod produced the current TradeCodex application during a window in which the Intent Engineer was also working on other projects, the AI Orchestrator was running four other stacks, and the Verification Owner was holding acceptance-contract discipline across the portfolio. Casey's share of the TradeCodex authoring was roughly five weeks of his own time, spread across the hundred days. The AI Orchestrator seat contributed about two weeks of my time. The Verification Owner seat contributed about a week of Glenn's time, most of it concentrated around the two or three VOSes where the acceptance contract needed architectural judgment rather than pattern matching.

Five plus two plus one. Eight pod-weeks, total, to produce what the traditional estimate pegs at thirty engineer-weeks. Multiplier against the traditional baseline for TradeCodex specifically: roughly 3.75× in pod-hours. That sounds low. It's low because the multiplier on pod-hours is not the multiplier the book publishes. The book publishes the multiplier against dedicated-engineer equivalents at the output level.

The thirty engineer-weeks at a single dedicated engineer works out to 150 engineer-days of concentrated work. The pod produced TradeCodex as a portion of the hundred days while also producing thirty-four other applications. Amortize the pod-hour cost across the portfolio and TradeCodex accounts for maybe six percent of the pod's sprint time. Six percent of one hundred days of pod time is six pod-days. One hundred fifty engineer-days divided by six pod-days of incremental cost is a 25× multiplier against the dedicated-engineer baseline, which is inside the CRUD web application range in the § 4 table (30–50×) but below the midpoint, which is honest for TradeCodex specifically because it has more financial-data edge cases than a typical CRUD app and therefore more acceptance-contract volume.

That's the shape of the math. Denominator: what a senior engineer would take to produce the same output at the same quality bar. Numerator: what the pod actually paid in incremental time to produce it. The ratio for any given application sits inside the work-type range the table shows. The blended portfolio ratio sits at 30–50× because the portfolio contains a lot of high-multiplier work (content sites, CRUD applications) and less low-multiplier work (native iOS, infrastructure) by LOC weight, even though the low-multiplier work consumed a disproportionate share of pod-hours.

If you want to cross-check the number yourself against a specific app, start with the LOC-per-application estimate in the at-a-glance summary, multiply by the traditional days-per-KLOC for a senior full-stack engineer (industry norm: roughly 4 to 8 engineer-days per thousand production lines at a shipping quality bar), divide by the pod-days we spent on that application. You'll land inside the range. If you land outside the range, we want to know, because either our counting is off or your baseline is.

---

5. What we counted

The multiplier rests on the counting. Here is the counting.

What we counted as production code. Application source in TypeScript, JavaScript, Swift, Python, and SQL; infrastructure authored in AWS CDK; tests written alongside the application code; generated scaffolding that the pod reviewed, named, and merged. If the code ran against real user data, or against the verification harness, or against deployed infrastructure, it counted. If a human would have been responsible for authoring and reviewing it under a traditional methodology, it counted here.

What we excluded. `node_modules`. Build artifacts (`dist`, `build`, compiled bundles, minified output). Scraped data files. Vendored dependencies. Package-lock.json and yarn.lock. Auto-generated files that neither a human nor the pod would have authored under any methodology (cached ASTs, generated type definitions, TypeScript build info). Test snapshots that exist as artifacts rather than as tests. Documentation (the methodology book you're holding has a lot of words in it; none of them are in the LOC count).

Included	Excluded
Application source (TypeScript, JavaScript, Swift, Python, SQL)	node_modules
Test files authored alongside application code	Build artifacts (dist, build, minified bundles)
Infrastructure as Code (AWS CDK)	Scraped data files
Generated scaffolding the pod reviewed and merged	Vendored dependencies
Shipped migrations	Package-lock.json, yarn.lock
Configuration authored for production	Auto-generated type definitions, cached ASTs
Acceptance contracts (Gherkin)	Test snapshots (artifacts, not authored tests)
	Documentation, marketing copy, book manuscript

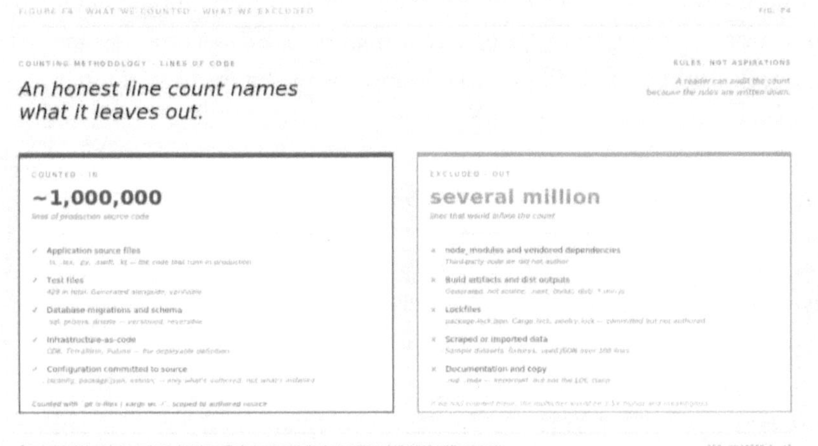

Figure F4 · Counted vs excluded LOC. *What the multiplier's numerator includes.*

The rough totals under the rules above: roughly one million lines of production source code across the thirty-five projects. About 19,500 functions and components. 429 test files. Two thousand commits against the repositories during the hundred days. Every repository is under version control and every commit is attributable to a specific author on the pod. A third party auditing the number would need access to the repositories, which is a conversation we're willing to have under an NDA with any acquiring publisher's technical reviewer.

I want to name one specific counting decision, because it's the one most likely to be challenged. We counted generated scaffolding that the pod reviewed and merged. Some readers will argue that generated scaffolding shouldn't count, because a human didn't type it. Our position is that a human authored the specification the scaffolding was generated from, a human reviewed the output, a human signed the acceptance contract, and a human merged the pull request. The fact that a machine produced the intermediate artifact doesn't change what the pod is accountable for. Same standard applies to a senior engineer who runs a code-generation CLI against a schema and reviews the output. The industry has never excluded that scaffolding from LOC counts. We're not going to exclude it either.

A second counting decision. We didn't inflate by counting AI dialogue as work. Conversations with the generation layer, failed attempts, prompts that produced nothing we shipped, these are not in the number. Only merged, shipped code counts. The methodology produces a lot of output that doesn't ship, the same way a traditional engineering team produces a lot of keystrokes that don't ship. What ships is the honest measure.

A third counting decision. The traditional-effort baseline is 9,150 developer-days (about 31 developer-years) based on industry-standard productivity benchmarks for senior full-stack developers. That's the denominator for the 30–50× multiplier. If you have a different baseline you trust more, you can divide our numerator by yours and get your own ratio. The numerator is the public fact. The denominator is the industry assumption.

If all of that feels pedantic, it's supposed to. The number gets challenged in every room it enters. We'd rather the book contained the chapter that tells you why the challenge doesn't land than produce a glossy headline and leave you to defend it.

---

6. The defect-rate claim, handled with discipline

I'm going to keep this section short on purpose, and the shortness is the argument.

We measured defect rates internally across the thirty-five-project portfolio. To compare those numbers against anything outside ourselves, the book needs a citable industry baseline. The trade-press "15 bugs per KLOC" figure most commonly quoted traces to sources we weren't willing to put our name next to. The baseline has to be one a skeptical reviewer can walk into the room holding. We landed on one.

The canonical published source is Steve McConnell, *Code Complete*, 2nd ed. (Microsoft Press, 2004) [1]. In the software-quality-landscape discussion, McConnell documents an industry range of roughly **fifteen to fifty errors per thousand lines of delivered code**, depending on the development practice in play. Structured programming with some level of discipline sits at the low end. General industry practice sits in the middle. Less-disciplined environments sit at the high end. McConnell's range has held up for two decades, is cited across the software-engineering literature, and is the number a serious reviewer will hold us to. Capers Jones publishes a parallel figure normalized to function points (roughly four defects per function point in total potential, with the delivered rate depending on removal efficiency) in *Applied Software Measurement* (3rd ed., McGraw-Hill, 2008) [2]; readers who want the function-point normalization can follow that trail. The two sources converge on the same order-of-magnitude claim. Mainstream human-written code delivers with double-digit defect densities per KLOC under ordinary conditions.

Here is what the book will and will not do with that baseline.

The book cites McConnell's fifteen-to-fifty-per-KLOC range as the industry baseline for delivered defect density in human-written code. That is the benchmark any comparison has to run against. Alchemaize's measured rate across the hundred-day sprint sits under three defects per thousand lines, which is below the lower end of McConnell's range and roughly five times better than the fifteen-per-KLOC midpoint a serious reviewer is most likely to anchor on. The discipline that goes with publishing the number is to publish what it counted, on what sample, in what window. The defect rate above is measured across our production repositories from the thirty-five-project portfolio, over the hundred days this chapter covers, against a definition that counts any production incident triaged as a code defect plus any post-deploy bug found in the first thirty days against the shipping VOS. A companion piece from Glenn, published outside the book and after it ships, walks through the measurement methodology in the depth a peer reviewer needs. Until that companion piece lands, the right way to read our number is as an internally measured rate that is consistent with the McConnell-anchored claim, not as a peer-reviewed benchmark. The methodology's upstream-verification posture is the reason the rate sits where it sits, and Chapter 6 covers the posture in detail.

This is the section of the chapter where restraint is the point. A book that claims five-times-fewer-bugs-than-industry without a citable source and without disclosed measurement is a book a skeptical reviewer walks away from. A book that cites McConnell correctly, publishes its own number with the disclosures attached, and points to a forthcoming companion piece for the full methodology earns the reviewer's attention instead of losing it.

One more source belongs in this section, as the forward pointer to Chapter 10. Forsgren, Humble, and Kim's *Accelerate* [3] established, through four years of survey data across thousands of engineering organizations, that specific delivery capabilities (lead time, deployment frequency, change-fail rate, time to restore) correlate with organizational outcomes in ways that ceremony adoption does not. The four numbers Chapter 10 proposes for a CID pod are not a new invention. They are the DORA capabilities, restated for a team whose coordination overhead has collapsed. The defect-rate claim in this chapter is the quality half of the argument. The throughput capabilities *Accelerate* measured are the speed half. Chapter 10 assembles both into the monthly meeting.

That is the work this section is willing to do in this edition. The rest comes later.

---

7. Casey as co-lead, the view from inside

Casey Robinson, Intent Engineer

The number is the part of this chapter David is best positioned to defend, because the methodology and the measurement were a testament to what we were creating. What I can tell you, from where I was sitting, is what the inside of the number looks like. Not the chart. The room.

I was in my office in Claremore most days. Light from the window, coffee to the left of the laptop, three screens showing different bits of information while my phone charges in front of me working through a VOS. That's the physical setup the number came out of for my seat. Whatever mental picture you have of how thirty-five applications got built, put mine in there with David's home office in Brentwood and Glenn's in College Station. Three rooms. Three screens. No physical building where we sat at all day.

The first few days on The Trade Codex were the training ground David mentioned. The first VOS took me about four hours to write. Why? Well it was the first time I really took a change to start something that was development adjacent and I wanted to read everything on the screen related. Going back and forth, reading what I was seeing on the screen, massaging the intent, making sure I got my Gerkin contract right, making sure I knew the outcome before it gets built. It was an exciting but also nerve racking time for me.

A VOS asks you to say what you want with a precision that would look too direct, or to much to the point where there is no extra interpretation.

That's the part nobody writing about AI coding has been honest enough about yet. The specification sentence for the dashboard feature I shipped on The Trade Codex looked like this: *The landing page after login sets the tone for the entire session. The dashboard surfaces the most important information: recent trades, active quests, P&L summary, market conditions, and Seer signals. It's the command center that routes users to deeper features. A dashboard that does not show the recent trades shall not be marred complete*

The old me before creating VOS wouldn't send that across in slack. I would add a bunch of words to try to explain it or shorten it to just one generic sentence. Which often times created questions if I did so. I would historically phrase it as *a trading dashboard to see everything trade related for a user.* An engineer reading that phrasing would fill in the gaps and try to do the right thing to interpret what we need or ask 10 follow on questions to draw out the information we needed to get it right. A machine reading that phrasing produces something close but misses the true intent of what I am looking for, and sometimes it's right and sometimes it isn't, and when it isn't, the bug is my fault, because I didn't say what I wanted.

Writing a VOS is the practice of saying what you want. That's the whole skill.

Here is what is interesting about writing VOSes. Once you have it written, you don't have to wait to go and implement it. The roles of the POD of CID are not tied to any one individual. After you right a VOS, you can leave it there to be picked up by a teammate, or you can run it right there as the AI orchestrator. There is no waiting after approval, no waiting for someone else to go and deliver, it is already there detailed out with your specifics, with history captured as part of the ELCID process to deliver.

So once you feel good with the VOS, have all the key VOS elements around the WHY, WHAT, HOW, CONTEXT, and OUTCOME it is now time to determine what comes next. I will save a deep dive into those aspects of implementation for another time, but know this, capture your intent with specifics and then put it together for your AI assistant to help fill in the gaps and review as the Human in the Loop before starting it to make sure the intent is clear.

One last note too, Don't think you just have to write the VOS then hand it off to someone else, the roles identified in CID are functions not people so be brave to go and implement after you have understood the intent of what is being built. The key is always making sure someone has independent verification of what is built and verified before moving forward with a scale out to broad set of users.

An early Trade Codex beat

Let me also share with you some aspects from those first days, because David's opening earlier in the chapter made it sound smoother than it was.

It was the end of January and I was really working hard through The Trade Codex, bringing the vision of the application to life. I'd written a VOS for the trade journal, the foundational feature of the whole app. The WHY was three sentences: *The core product: a trade journal where users log stock and options trades, track P&L, view history, and get AI-powered analysis. Supports multi-leg options (verticals, iron condors), scaling entries/exits, and calendar views. The journal is the foundation everything else builds on.*

The acceptance contract was in Gherkin. About fourteen lines. Log a trade, it shows up in the journal and the calendar. Multi-leg options trades store individual strikes, expirations, and premiums with combined P&L across legs. Scaling entries and exits calculate average prices and track partial positions. Filters for date range, strategy, ticker, status. Calendar view with daily P&L summaries. Trade detail page with chart and AI analysis.

I wrote it, ran it through the generation layer, and got back the journal, tests, and what looked like reasonable code. Tests passed. Acceptance contract passed on the first run. I was a few days into the methodology and I thought I'd gotten good at it faster than David had told me I would.

Glenn reviewed the VOS the next morning and flagged something I'd missed. Not the code. The code was fine. He'd read the multi-leg options scenario and identified a gap in the acceptance contract. My contract didn't account for what happens when a user scales into a spread position over multiple entries. The journal would show the trade. The P&L would calculate. But if someone entered the first leg on Monday and the second leg on Tuesday, the combined position wouldn't reconcile correctly because my contract didn't specify how partial multi-leg positions should display before all legs were filled. The user would see a number. The number would be off. And nothing in my contract would catch it.

His perspective on it was stronger than mine. The specification had passed in the sense of the machine's work, but it had fallen short in the sense of the user's outcome. The machine had done exactly what I'd asked. What I'd asked was incomplete.

I improved the contract based on Glenn's feedback. Added three clauses. Two of them Glenn was good with immediately. The third we went back and forth on, because I felt his version over-specified a case that wouldn't show up in real data. After we talked it through and ran the contract against real trade data we had sitting in a test bucket from TradeCodex's early users, the case Glenn had flagged surfaced in the second file. I came around to his perspective and we went with his version.

The generation layer produced a new journal in about four minutes. Tests passed. Contract passed. Shipped.

That's the shape of the work I want you to see. The generation layer was never the problem. The contract was the problem. My contract. The part I hadn't been thorough enough about. Glenn's review caught it because he brings a career architect's perspective, with two patents in cryptographic systems and a mental model of where silent failures hide. My perspective couldn't have caught it at the end of January. It can now, because I've written enough contracts to have built some of that mental model myself. The methodology is a training ground as much as it is a delivery system.

One thing I want to be clear about here. The roles in CID, the Intent Engineer, the AI Orchestrator, and the Verification Owner, are functions, not people. On that first Trade Codex VOS, I wrote the intent, ran it through the generation layer myself, and Glenn reviewed the output. I played two of the three functions. On later VOSes I'd play all three and then hand it to Glenn or David for independent verification, which is the ELCID requirement. The methodology doesn't require a handoff between three different people. It requires that someone who didn't build the thing reviews the thing before it goes to users. How you get to that review is up to the pod.

David didn't share this particular story earlier in the chapter, so I wanted to make sure it was part of the picture. It's part of the hundred days, and it's the kind of moment that shaped how I think about writing contracts today.

The VOS I almost threw away

I want to tell you about one specific VOS, because I think the chapter is pointed at the wrong number and this VOS is what I'd point at instead.

It was a Saturday in mid-February, and I was sitting in my office in Claremore working on The Trade Codex. I'd been at it for a couple of hours already, deep into the Campaign & Quest System, which was VOS #4 in the project. This was one of the features I was most excited about. The idea was to wrap trading education in a narrative-driven RPG experience with campaigns, levels, side quests, XP, badges, and a world bible. The first campaign, Initiate's Ascension, would teach fundamentals through story. The WHY was two sentences: *Learning to trade is hard and lonely. The campaign system wraps trading education in a narrative-driven RPG experience with campaigns, levels, side quests, XP, badges, and a world bible.*

The acceptance contract was about twenty lines of Gherkin. Play a campaign level, complete a quest by logging a specific type of trade, earn badges, answer questions from an AI-generated question bank, progress through the campaign and unlock the next one. The HOW section had seven tasks. The CONTEXT bundle had the campaign module, the quest system files, the database schema for progress tracking, and the steering docs for the world bible.

I was reading it one more time before I fired it at the generation layer. Something felt off. I couldn't name it at first. The contracts were specific. The context was tight. The outcome made sense. But the whole thing felt too big.

The VOS bundled two behaviors that didn't really need each other. First, the campaign progression system: levels, XP, badges, the narrative flow, the unlock logic. Second, the quest validation system: the part where a quest says, "log a vertical spread trade" and the AI validator checks whether the user actually did it. I'd written them as one unit because in my head they were one feature. From the user's perspective they probably are. But the acceptance contract for each behavior didn't interact with the acceptance contract for the other.

The campaign-progression contract is about the UI, the level flow, the XP math, the badge awards. The quest-validation contract is about the AI checking real trade data against quest objectives, the Bedrock integration for validation, the edge cases where a user logs something close but not quite right. Two different contracts. Two different kinds of failure mode. Two different things I'd want reviewed separately.

I spent about forty minutes cutting the VOS in half. First half: campaign progression with its own acceptance contract. Second half: quest validation with its own acceptance contract. The context bundles changed. The campaign VOS pulled the level data and the UI components. The quest VOS pulled the trade journal integration and the Bedrock validator. Neither Glenn nor David had asked me to split it. I made the call on my own because reading my own specification one more time told me the split was cleaner.

I shipped the campaign progression that Sunday. Ran it through generation myself, verified the output looked right, then handed it to David for independent review. He was good with both contracts on the first look. Shipped the quest validation Monday afternoon. This time Glenn reviewed it and flagged one of the contracts. He'd spotted a case where the AI validator would approve a quest completion based on a paper trade instead of a real one, which would let someone game the progression without actually trading.

His perspective made a lot of sense once I saw it. Fix took twenty minutes. Shipped.

Here's why I'm telling you this.

The thirty-five-app number is the headline, and it's real. I helped produce it. But if you ask me what the chapter should be pointed at, I don't think it's thirty-five. I think it's that on several of those applications, I was authoring specifications and running them end-to-end, from intent through generation to independent review, against codebases where the deepest domain knowledge in the room was mine. And mine is not a senior engineer's perspective. It's a twenty-year operator's perspective.

That worked. The VOS I almost threw away worked, and it worked because the methodology gave me a way to catch my own gap before the machine amplified it. Splitting the VOS wasn't a technical call. It was a clarity call. I could make it because the methodology is built around clarity being the work.

If the chapter's claim is thirty-five in a hundred days, fine, that's the claim. My claim, the one I'd want a reader to hold as they go into the rest of the book, is: the pod shipped production software where someone playing the Intent Engineer function went end-to-end on the build and independent verification was the only step between the work and the user. The bottleneck moved far enough that the function I was filling could produce output the old industry said someone in my position could not produce. That's the thing I think readers should be arguing about.

The chart is not the headline. The person who filled the chart is the headline.

I just wanted to share a little bit from my perspective on what those hundred days looked like from the inside. Now let me hand it back over to David to continue the journey here in the chapter.

---

8. What the number does not prove

Casey is partly right and partly sharpening me in public, which is the correct division of labor on a co-lead chapter, so I'll leave his argument standing and add to it rather than rebut it.

The hundred-day number proves a specific thing. It proves that a three-person team with one non-engineer, using the methodology this book describes, operating distributed across three states with no shared physical workspace, over a fourteen-week window beginning January 18, 2026, shipped thirty-five production applications totaling roughly one million lines of production source code, under a counting methodology disclosed in § 5, against a traditional-effort baseline of approximately 31 developer-years, producing a blended 30–50× throughput multiplier. The window was the second 100 days, not Alchemaize's first; the first 100 days produced an MVP and a lesson, and the second 100 days are what the lesson made possible. Every word of that sentence carries a specific load. Read it again if you need to.

Here is what the number does not prove.

It does not prove the methodology scales to a hundred engineers. The pod is three people. The ELCID layer in Part III argues the methodology scales, but the argument there is architectural, not empirical, and it deserves to be read as architecture. The book you'd write to prove scale empirically is a different book with a different window and a different dataset. We're honest about that.

It does not prove the methodology works in a regulated industry without additional verification architecture. TradeCodex handles financial trading data but is not itself operating under HIPAA, SOX, or any regulated-industry compliance regime that would change the verification posture. Chapter 11 addresses compliance specifically and argues a position some readers will find surprising. The argument there doesn't rest on the thirty-five-app ledger; it rests on the structural properties of upstream verification under AI-native generation.

It does not prove the methodology works for a team without the three-author configuration we ran. Casey's argument in § 7 is that the pod operated without a senior engineer on four of the projects, and that's true, and it matters. But the pod had Glenn in the verification seat across every one of the thirty-five. Glenn's architectural depth is the scaffolding the verification seat sat on. A pod without a Glenn-shaped Verification Owner is a different experiment. We believe the role is replicable. We haven't yet run the experiment that proves it. Chapter 6 is the chapter that tries to name what a Verification Owner needs to be and how you build one who isn't Glenn.

It does not prove that every organization can produce these numbers by adopting the methodology. The Alchemaize pod had specific advantages. Twenty years of working trust between two of the three founders. A small surface area and no legacy code to drag. A founding team that had all three of them at AWS or AWS-adjacent, which meant the shared vocabulary came for free. Organizations adopting CID from inside a large enterprise will hit friction we didn't hit. Chapter 13 names that friction and the adoption path we recommend. The thirty-five-app number is the ceiling we reached in a green-field setup with an aligned founding team. What it produces for you is going to depend on your starting conditions.

What the number does prove, inside its scope, is that the methodology can produce what it claims to produce. That's the structural job of an evidence chapter. Not to prove every future inference. To prove the base case the inferences attach to. The inferences attach in the next eleven chapters.

You've seen the ledger. You've seen the counting. You've seen the multiplier three ways. You've heard from the person who held the Intent Engineer seat for most of the hundred days. You've heard what we won't claim and why. If you're still with the book, the rest of the argument rests on a floor you can see. If you're not, you at least know which floorboard broke, and that's a useful thing to leave a reader with.

The evidence is on the table. The methodology is what it's evidence for.

Notes

1. Steve McConnell. *Code Complete,* 2nd ed. Microsoft Press, 2004.
2. Capers Jones. *Applied Software Measurement,* 3rd ed. McGraw-Hill, 2008.
3. Nicole Forsgren, Jez Humble, and Gene Kim. *Accelerate: The Science of Lean Software and DevOps.* IT Revolution, 2018.

Cross-references: Chapter 1 (the Shift) sets up the bottleneck argument the multiplier rests on. Chapter 4 (the VOS) defines the atomic unit this chapter counts. Chapter 6 (verification) is where the counting discipline § 5 describes gets its architectural treatment. Chapter 8 (Casey's chapter) expands the Intent Engineer seat the Field Note sketches. Chapter 13 (adoption) addresses the starting-conditions question § 8 raises.

PART II

The Practice

Chapters 4 through 8.

The methodology itself. The atomic unit, the context discipline, the verification posture, the pipeline in motion, and the chapter the methodology was built for. By the end of Part II, you have what a pod actually runs.

CHAPTER 4: INTENT AS THE ARTIFACT

For three chapters I've been walking around the diagram without drawing it. That was deliberate. A picture on a page is easy to nod at and easier to forget. A picture on a page that answers a question you've already been holding for forty pages is the kind of image that sits on the desk the rest of the year.

Here's the question, restated so we know what the image is answering. If typing is no longer the expensive step, what is the organization building around? If the frameworks that coordinated typing are friction, what replaces them? If a non-engineer can ship production code under the right framework, what is that framework doing that lets him?

The answer is a pipeline. Four stages of human and machine work running in sequence, plus a watching layer that runs in parallel. Intent. Context. Generation. Verification. The watching layer is Observation. The pipeline is per-VOS, forward-flow. The watching layer runs continuously across every VOS already shipped. Together, they describe what has to happen, in what order, and what watches it from the side, for a piece of software to go from a business need to a running feature.

The pipeline has a name. It's the one on the cover of the book. Continuous Intent Delivery. What you've been reading is the argument for why the pipeline exists. What this chapter does is draw the pipeline and walk through the piece of paper that travels through it.

1. The inversion

Let me say the thing the pipeline is built to say, in the fewest words it'll take.

Intent is the artifact. Code is the exhaust.

A career engineer reading that sentence has two reactions at once. The first is discomfort. Code has been the artifact for fifty years; it's what ships, what compiles, what runs, what the job was. The second reaction, sitting underneath the first, is a quieter recognition that maybe the discomfort is evidence rather than objection. The inversion reorganizes where the creative act lives. It doesn't abolish the creative act; it relocates it.

Here is what the sentence means. The artifact an engineering organization should author, version, review, argue over, and keep as its durable knowledge is the intent specification. The code is what falls out when the specification is right. A well-run pipeline treats the code as the output of a process, not the input to one, and within reasonable bounds the organization can regenerate the code from the intent. The intent is the thing that has to survive.

That's a claim about where humans should be spending their attention. Attention is the organization's most constrained resource. For twenty-five years it was spent on typing, on reviewing typing, on coordinating typing, on fixing the typing after it was done. Typing was the expensive middle of the pipeline, so the organization's attention had to go there. That was rational under the old constraint.

The new constraint asks for a different allocation. Specification and verification are now the dominant ends of the pipeline, with machine labor doing the generation step in between. Human attention is the right tool for what a machine cannot do on its own and the wrong tool for what a machine handles well. The pipeline looks for human attention at the two dominant ends: the intent that goes in, the verification that comes out.

The four pipeline stages and the watching layer follow from that design choice.

Intent. A human authors a specification of what the software should do, for whom, and to what measurable end. This is where the human thought lives. The specification is a written document, small enough to read in a sitting, specific enough to be acted on without a hallway conversation.

Context. A human curates the set of files, patterns, and prior decisions the generation layer needs to act correctly. Not the whole codebase; the relevant pieces. Context is a selection act, not a dump.

Generation. A machine produces code against the intent specification and the context bundle. This is the step that used to be weeks of typing. It is now minutes of inference.

Verification. The intent specification includes an executable acceptance contract. The contract runs against the generated code. If the contract passes, the work moves forward to SHIPPED. If it fails, the work goes back to the upstream stage that owns the gap, not forward to a fix-later bucket. There are two such reverse edges and SHIPPED is terminal.

Observation. This is the watching layer. It runs alongside the pipeline, continuously, against the population of VOSes already in production. The specification of every shipped VOS included a hypothesis. Observation is where each hypothesis meets reality. When observation surfaces actionable signal, that signal becomes the seed for a new VOS, drafted fresh, that takes its own one-way trip through the pipeline. Observation does not push the old VOS back through. The old VOS is shipped.

Four pipeline stages. One watching layer. Three of the pipeline stages are what humans do well. One is what machines do well. The watching layer is how the system learns from its own deployed work, at a system-level cadence the pipeline does not see. Every pipeline stage has an output the next stage consumes. The output of Intent is a specification. The output of Context is a bundle. The output of Generation is code plus tests. The output of Verification is a pass or a documented rejection. The output of the watching layer is a stream of new candidate intents, fed back into the head of the pipeline whenever the pod is ready to draft another VOS.

The hero diagram of the book sits opposite this spread.

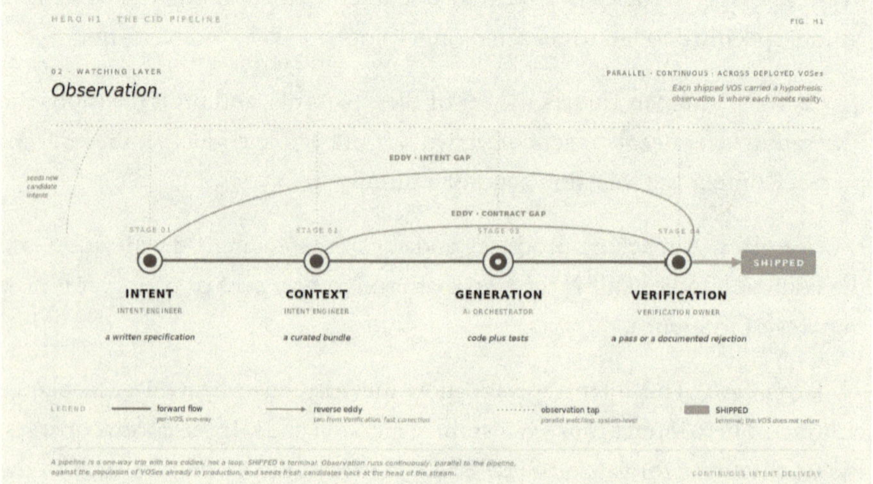

The CID Pipeline. *Four stages forward, two reverse edges, parallel watching layer.*

What the pipeline is not: a closed loop. SHIPPED is terminal. The two reverse edges from Verification are eddies inside one VOS's trip, not a feedback cycle that returns the same artifact to its origin. The watching layer connects production to intent at the system level asynchronously and stochastically, not the closed cycle PDCA or OODA describe. What the pipeline is also not: waterfall. Cycle time per VOS is hours or days, the eddies allow rapid contract-and-regenerate iteration, and the watching layer runs in parallel rather than as a final QA phase.

The artifact in the phrase "intent is the artifact" is a specific piece of paper with a specific shape that a specific person authors. You can hold it in your hand. It has five sections. We call it a VOS.

2. The VOS, section by section

Verifiable Outcome Slice. The acronym is ugly and it has stuck anyway. A VOS is a markdown file. It has one H1 title and five required H2 sections. A VOS describes exactly one unit of user-observable value. Not a sprint's worth, not a theme, not an epic. One unit. If the WHY paragraph takes longer than one paragraph to write, the slice is too big and you split it.

The five sections, in order, are WHY, WHAT, HOW, CONTEXT, and OUTCOME. Each one has a different author on the pod. That's not a decorative detail. The section-to-role mapping is part of what makes the methodology work.

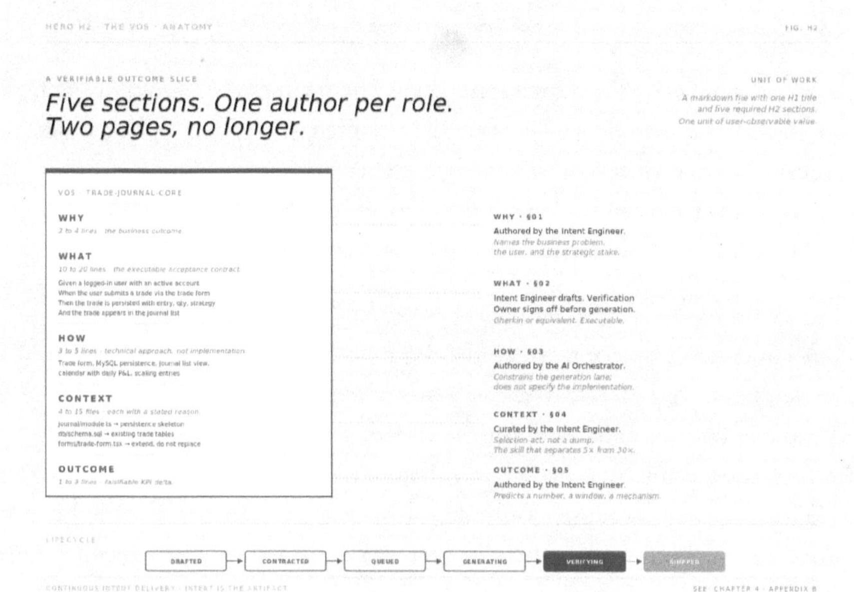

VOS Anatomy. *The five sections (WHY, WHAT, HOW, CONTEXT, OUTCOME)*

Casey wrote a VOS in late January for the foundational feature of The Trade Codex, the trading journal and education platform he was building. The feature was called Trade Journal Core. It is a fine example of a VOS because it is an ordinary one. The trade-journaling problem is straightforward. The VOS is well-formed. We will walk through it.

WHY

Two to four lines. Authored by the Intent Engineer. The WHY answers three questions: what business problem the slice solves, who benefits and how, and how the slice connects to the stream's strategic intent.

From Casey's VOS, verbatim:

> The core product: a trade journal where users log stock and options trades, track P&L, view history, and get AI-powered analysis. Supports multi-leg options (verticals, iron condors), scaling entries/exits, and calendar views. The journal is the foundation everything else builds on.

That's the whole WHY. Three sentences. The problem is named (users need to log and track trades). The scope is named (multi-leg options, scaling entries, calendar views). The strategic stakes are named (the journal is the foundation for everything else). No buzzwords. No hedging. The sentence would look rude in an email and it is exactly the register the document calls for.

A bad WHY, by contrast, reads like this: *Improve the trading experience and user engagement by providing a comprehensive journaling solution.* That sentence is unreadable by a machine and unreadable by a colleague. It answers none of the three questions. A specification that starts there will produce generated code that is, on average, journaling-ish, and sometimes it will be and sometimes it won't, and when it isn't the bug will be the author's fault because the author didn't say what they wanted.

The WHY is where the discipline starts. Every VOS that passes verification starts with a WHY that could survive a skeptical reader asking, "so what." The Intent Engineer has to be willing to commit to the so-what in two or three sentences, on a first draft, without a paragraph of throat-clearing before it.

WHAT

Ten to twenty lines, usually. Authored by the Intent Engineer with the Verification Owner as a required second reader. The WHAT is the acceptance contract, written in Gherkin, before any code exists.

Gherkin is a specification language that reads like English but is executable. It has three keywords that carry the work: Given, When, Then. Given sets up the state. When names the action. Then names the observable outcome. A Gherkin scenario is a small, testable claim about how the software should behave.

Here is part of the WHAT from Casey's VOS:

```
Feature: Trade journal
  Background:
    Given a logged-in user with an active Trade Codex account
  Scenario: Log a new trade
    When the user submits a new trade via the trade form
    Then the trade is persisted with entry price, quantity, strategy, and notes
    And the trade appears in the journal list and calendar
  Scenario: Multi-leg options trade
    Given the user is logging an options trade
    When they add multiple legs (e.g., vertical spread)
    Then all legs are stored with individual strikes, expirations, and premiums
    And combined P&L is calculated across legs
  Scenario: Trade history and filtering
    When the user views the journal
    Then trades are listed with status, P&L, strategy, and dates
    And filters are available for date range, strategy, ticker, and status
```

Three scenarios. Primary success path (log a trade), a complexity case (multi-leg options), and a usability case (filtering and history). The WHAT is the smallest complete description of what the feature has to do for a user to trust it.

The Gherkin looks simple. The discipline underneath it is not. The Intent Engineer has to know, in advance, what the software should do in the three or four or five scenarios that actually matter. The Verification Owner has to know, in advance, whether those scenarios are the right ones, whether the error cases are specified tightly enough, whether the edge cases will catch the bugs that will actually happen.

Most specifications fail at the WHAT. An Intent Engineer who writes one scenario (the happy path) and calls it done has under-specified the contract. The generated code will pass that one scenario and will fail every case the scenario didn't cover. The feedback, when it arrives, comes from the pod's review of the contract, not from a build error. Casey's Field Note at the end of this chapter goes deep on what learning to write a WHAT actually felt like.

HOW

Three to five lines. Authored by the AI Orchestrator, which on our pod is me. The HOW is not an implementation specification. It names the technical approach at the level of what systems get touched and what patterns get reused, not at the level of how any individual function gets written.

From Casey's VOS:

> Trade form with multi-leg support. Trade persistence in MySQL. Journal list view with filtering and sorting. Calendar view with daily P&L. Trade detail page with chart integration. Scaling entries/exits with average price calculation.

Four sentences. Two positive instructions, two negative ones. The negative instructions matter. The AI Orchestrator's job is to keep the generation layer inside the architectural lanes the team already owns. "Do not add new data-model tables" is a sentence that prevents a particular class of drift the generation layer is prone to, which is inventing a plausible-looking schema that makes the feature work locally but corrupts the data shape of the product over time.

The HOW is short on purpose. If the HOW section is longer than five lines, the AI Orchestrator is doing the generation layer's job instead of constraining it. The generation layer will figure out function signatures, variable names, and internal control flow on its own. What it will not figure out on its own is which parts of the existing codebase it is allowed to touch, which patterns it is required to reuse, and which tempting shortcuts it is forbidden to take.

A good HOW fits on a single screen. A great HOW fits in a single paragraph. If I catch myself writing pseudocode in the HOW section, I stop and delete the pseudocode. That's a sign I've drifted into the generation layer's work.

CONTEXT

This is where Casey picks up the pen.

> *The CONTEXT section is the one I've gotten the most wrong, the most often, and the one where I've gotten the most better.*
>
> The CONTEXT is a list of files. Not a description of the whole application. Not everything. A list. Each file on the list has a short reason next to it that says why the generation layer needs to see it. I use my AI tool to help me identify the specific files once I describe the feature areas in plain language.
>
> For the RFP analysis VOS on BidForge, my CONTEXT was four areas. The document parsing service, because the analysis feature has to read the extracted requirements from the uploaded RFP. The capability repository, because the AI needs to match requirements against what the organization can actually do. The scoring engine, because the fit rating for each requirement feeds into the composite score. The opportunity detail page, because that's where the analysis results show up for the user.

That's it. Four areas. BidForge has a lot of files in it. Only four areas belonged in the context bundle for this VOS, and my AI tool helped me identify the specific files from those descriptions.

The instinct, when you're new at this, is to include more. You think you're helping. You think more context has to be better context. It isn't. The generation layer will use what you give it. If you hand it fifteen files, it will try to use fifteen files, and a bunch of them will be off, and the code it produces will be a reasonable response to the fifteen-file bundle you handed it rather than to the problem you were trying to solve.

I've described this to people as setting the table. You're not handing the machine the kitchen. You're handing it a specific set of ingredients for a specific dish. Too many ingredients and the dish is a mess. Too few and the dish isn't complete.

The right number of files is almost always between four and fifteen. If I'm under four, I'm probably missing something and will find out on the first generation pass. If I'm over fifteen, I'm almost always including areas that I thought were important for me to understand but that the generation layer doesn't actually need. That's a different job. What I need to understand about the application is not the same as what the generation layer needs to produce correct output. I have to separate those two things.

There's a skill in the selection that I didn't have in January and that I'd started to have by March. The skill is knowing which features and areas of the application are relevant to the VOS and which ones aren't. My AI tool helps me translate that understanding into the actual files, but the judgment about what to include comes from understanding the product at the feature level.

I spend more time on CONTEXT than on any other section of a VOS. WHY is usually thirty minutes. WHAT is usually an hour. HOW is usually ten minutes because David writes most of it. CONTEXT is often two hours. It's the section where the VOS earns its quality on the first pass or doesn't.

Casey is describing the section that, in the old methodology, didn't exist. The user story format Schwaber and Sutherland's *Scrum Guide* [1] defines (*as a user, I want X, so that Y*) did not name files because the engineer reading the story already knew the codebase from memory. The requirements document was a rough sketch and the engineer filled in the rest from memory. That filling-in step has now moved onto the page, because the generation layer, for all its capability, has no memory of your codebase until you hand it one.

The Intent Engineer does that work. Casey does it for Alchemaize. The observation he is too polite to state directly, which is that a non-engineer is often better at this particular section than the engineers around him, is true. A person who has spent years reading codebases rather than writing them has a selection instinct the writers don't always have. Those are different skills and the second one is what context curation rewards.

OUTCOME

One to three lines. Authored by the Intent Engineer.

From Casey's VOS:

> Reconciliation-failure rate visible to users drops to zero over the next billing cycle, because any projection we can't reconcile we won't show.

One sentence, one number (zero), one measurement window (next billing cycle), one mechanism (we won't show what we can't reconcile). The OUTCOME names a testable claim about the world. The VOS isn't done when the code ships. The VOS is done when the outcome has or hasn't happened.

Casey will tell you that writing a good OUTCOME was the hardest habit to build. The temptation is to write something aspirational and fuzzy (*improve customer trust in the product*) because aspirational-and-fuzzy is unfalsifiable, and unfalsifiable is safe. The discipline of the OUTCOME section is that it has to be falsifiable. The slice is predicting something. The organization will know in six weeks whether the prediction came true.

> *I'll add one thing to what David said about the OUTCOME.*
>
> The first OUTCOMEs I wrote were all aspirational. I'd write something like, "the reconciliation workflow feels more trustworthy to users." Glenn would read it and send it back with a one-line note. *What would you measure. What number would tell you this was true.*
>
> So I'd go back and try again. My second version would be, "fewer support tickets related to reconciliation." Glenn would send that one back too. *How many fewer. Over what window. What's the current number.*
>
> Third version was, "reconciliation-related support tickets drop by at least half, within 30 days of ship, from a current baseline of roughly twelve per week." That one he accepted.
>
> The pattern I noticed is that every time Glenn rejected an OUTCOME, he was right, and he was right for the same reason. An OUTCOME that can't be falsified isn't an OUTCOME. It's a wish. Wishes don't make a falsifiable claim. Falsifiable predictions do. You write down what you think will happen, you ship the thing, you watch the number, and the number tells you whether you were right.
>
> The version in the reconciliation VOS, the one David quoted, I got on the first try. That's what three months of practice looks like.

Five sections. A VOS fits on two to three printed pages. It describes one unit of user-observable value, names an author on each section, and is the thing humans author, version, review, and argue over. It is, in the phrase this chapter is built around, the artifact.

Five years from now, the Trade Codex code may have been rewritten three times. The VOSes that shipped the original features will still be the record of what was intended, what was verified, and what outcome was predicted. That's what durability looks like when intent is the artifact.

3. The six lifecycle states

The lifecycle is a directed graph with two reverse edges, not a cycle. SHIPPED is terminal. A VOS enters at DRAFTED, moves forward through the states, and exits the graph when the Verification Owner signs. The two reverse edges from VERIFYING exist for the cases where the generation missed or the contract was wrong, and they let the pod recover cheaply without breaking the forward shape.

The states are visible. They live in the filesystem. A VOS in a particular state sits in a particular folder. State transitions are committed to git. The filesystem is the truth.

The six states are DRAFTED, CONTRACTED, QUEUED, GENERATING, VERIFYING, and SHIPPED. Each transition has a specific trigger.

DRAFTED. The Intent Engineer has written the WHY and a first pass at the WHAT. The AI Orchestrator has not yet reviewed. The Verification Owner has not yet signed. The file lives in a drafted folder.

CONTRACTED. The WHAT has been reviewed and signed by the Verification Owner. The acceptance contract is now binding. The HOW section has been written. The CONTEXT bundle is complete. The OUTCOME is stated. The VOS moves to a todo folder.

QUEUED. The pod has scheduled the VOS for generation. Not all contracted VOSes are in the queue at once. Queueing is how the pod controls work-in-progress.

GENERATING. The generation layer is producing code against the VOS. The VOS sits in an in-progress folder. The code and tests are being produced. The AI Orchestrator is watching.

VERIFYING. Generation is complete. The acceptance contract is being run. The Verification Owner is reviewing. If the contract passes and the reviewer agrees the outcome matches the intent, the VOS proceeds. If not, it goes backward.

SHIPPED. The code is in production. The VOS sits in a shipped folder. The VOS is now immutable. It is part of the organization's durable record of what was built and why.

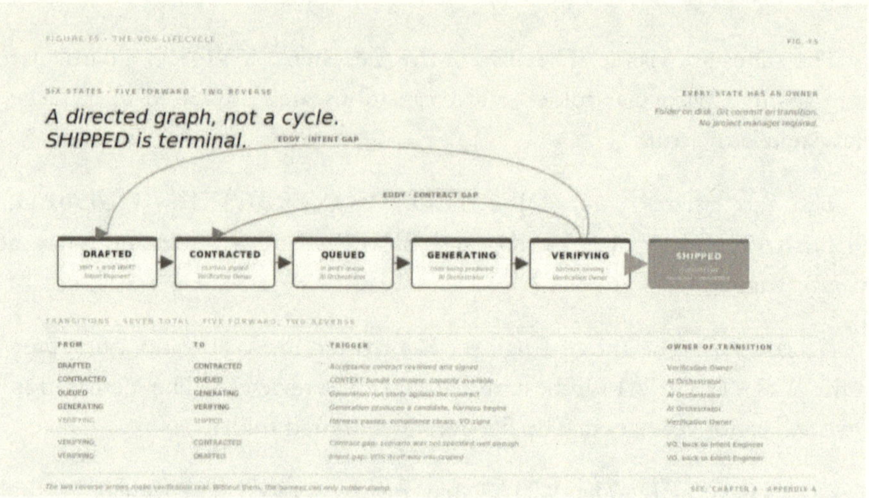

Figure F5 · VOS lifecycle state machine. *Six states (DRAFTED, CONTRACTED, QUEUED, GENERATING, VERIFYING, SHIPPED)*

Glenn writes:

The reverse arrows are what makes verification real.

If a methodology only lets verification move a VOS forward, then verification is mostly theater. The verifier does not really have authority. They can approve the work, or they can hold up the line, which over time becomes its own form of approval, because the pressure to ship does not go away and the verifier is the only thing standing in front of it. That is not real verification. That is a rubber stamp with a slow arm.

In CID, verification has two reverse paths, and both matter.

If a VOS moves from VERIFYING back to CONTRACTED, it means the acceptance contract was wrong, unclear, or incomplete. The work surfaced something the contract did not define well enough up front. The right answer is not to patch around it. The contract gets rewritten before the pod tries again.

If a VOS moves from VERIFYING back to DRAFTED, the issue is deeper. The VOS itself was off. The original intent did not hold up once the team got into the work, and the pod needs to step back and rethink what it was actually trying to accomplish.

The first case is a contract failure. The second is an intent failure. Both are normal, and both will happen. The important thing is that both must be inexpensive to recover from. If sending work backward feels costly, teams will quietly stop doing it. They will force things forward instead, and verification will slowly turn back into the rubber stamp it was supposed to replace.

The mechanic itself is simple. When a VOS fails verification, it goes back to the state where the real problem belongs. Not forward into a generic fix-it bucket. Not sideways into a defect queue. Backward, to the right place in the pipeline, so the pod can restart from the correct stage and keep moving.

That is what keeps the acceptance contract honest.

The table below names every transition, its trigger, and its owner. Five forward, two backward. The pod can look at any VOS on any day and see what state it's in and who has the next move.

From	To	Trigger	Owner of the transition
(new)	DRAFTED	Intent Engineer writes WHY, WHAT, HOW, CONTEXT, OUTCOME	Intent Engineer
DRAFTED	CONTRACTED	Acceptance contract reviewed, all scenarios verifiable, Verification	Verification Owner

		Owner signs the contract rubric	
CONTRACTED	QUEUED	Context bundle complete, stream capacity available	AI Orchestrator
QUEUED	GENERATING	Orchestrator starts the generation run against the contract	AI Orchestrator
GENERATING	VERIFYING	Generation run produces a candidate, harness run begins	AI Orchestrator
VERIFYING	SHIPPED	Harness passes, compliance extensions clear, Verification Owner signs	Verification Owner
VERIFYING	CONTRACTED	Contract gap: the scenario the harness failed on was not specified well enough to regenerate against	Verification Owner, back to Intent Engineer
VERIFYING	DRAFTED	Intent gap: the VOS itself was mis-scoped, the intent didn't survive contact with the work	Verification Owner, back to Intent Engineer

Table T3 · Six states, seven transitions.

There is no ticketing system adjacent to this. The filesystem and git are the ticketing system. The VOS file in its folder is the ticket. The git history of the VOS file is the audit log. The transition from one folder to another, committed to the repo, is the state change. A project manager who wants to know the status of the pod looks at the repo. A verifier who wants to review a contract looks at the file. A new team member who wants to understand what the team shipped in March reads the files in the shipped folder. The work and the record of the work are the same thing.

4. The seven first principles

#	Principle	What it replaces

1	Intent is the artifact. Code is the exhaust.	Code as the durable record
2	Verification is the trust anchor.	Downstream QA as a safety net
3	Flow over cadence.	Sprints, release trains, calendar-driven delivery
4	Fund streams, not projects.	Project-based budgeting and fixed-scope plans
5	Three roles, not thirty.	The expanded role chart of SAFe and its descendants
6	Measure outcomes, not activity.	Velocity, story points, ceremony attendance
7	Ceremony is a tax.	Standups, sprint reviews, retrospectives, PI planning

Table T4 · The seven First Principles. Each principle is expanded in the prose that follows; the rest of the book refers to them by number.

The rest of the book will refer to these by number and by name. Here they are, stated once, briefly, as the spine that holds the chapters together.

1. Intent is the artifact. Code is the exhaust. The thing humans author, version, review, and argue over is the intent specification. The code is the predictable output of a well-run pipeline. This is the inversion the chapter opened with. Everything else in the methodology follows from it.

2. Verification is the trust anchor. The acceptance contract is written before the code and is executable. A VOS cannot ship until the contract passes. The verifier has the authority to send work backward through the pipeline. Without this, the rest of the methodology is a prompt-engineering exercise with nicer handwriting.

3. Flow over cadence. Work moves when it's ready, not when a two-week calendar says it's time. The sprint is a coordination artifact from the typing era. The pod runs continuous flow against a stream of intent. What the sprint boundary was for, in the old world, is handled by the verification gate in the new one.

4. Fund streams, not projects. A project is a fixed-scope, fixed-schedule, fixed-budget instrument from a world where typing was the constraint. A stream is an intent with an outcome hypothesis and a pod funded to pursue it. Streams fund intents, not scopes. Chapter 9 expands this into the enterprise layer.

5. Three roles, not thirty. The pod is three people: an Intent Engineer, an AI Orchestrator, and a Verification Owner. There is no separate product manager, tech lead, quality engineer, or scrum master. The three roles cover the pipeline, and more people dilute the roles and re-introduce the coordination cost the methodology is designed to remove. The pod is a stream-aligned team in the sense Matthew Skelton and Manuel Pais use in *Team Topologies* [2]: a single, durable team that owns a slice of the business end-to-end, sized so cognitive load stays manageable. The CID pod takes that shape and specifies the three roles it requires.

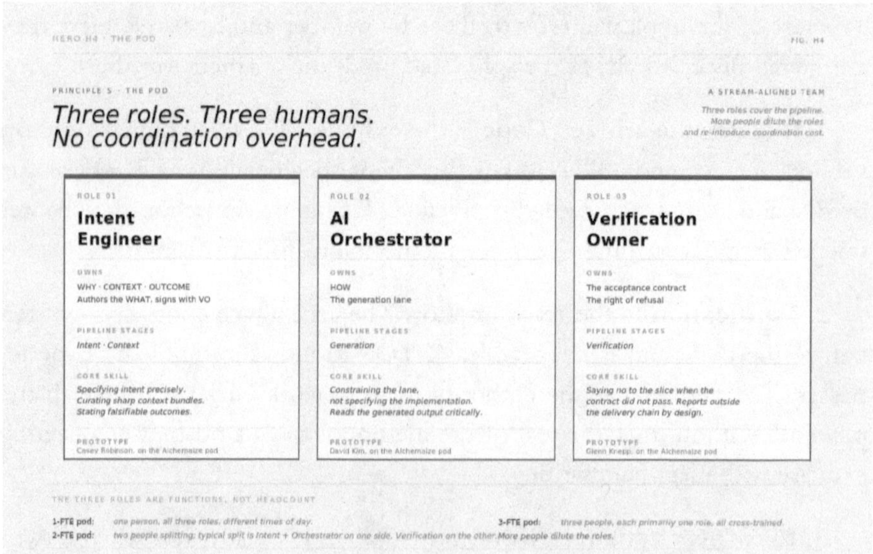

The Pod. *Three roles, three humans: Intent Engineer, AI Orchestrator, Verification Owner.*

6. Measure outcomes, not activity. Velocity and story points are typing-era metrics. What you measure in CID is whether the OUTCOME a

VOS predicted actually happened, and whether the pod's VOSes are shipping with high first-pass verification rates. Two numbers, roughly. Chapter 10 goes into more depth.

7. Ceremony is a tax. Every meeting, artifact, or process step that doesn't directly move intent through the pipeline is a tax on the pod's ability to ship. The methodology starts from zero ceremony and adds back only what the pod can justify. Standups, sprint reviews, and PI planning are taxes the pod pays out of the same attention budget it uses to think about intent, and that budget is finite.

Seven principles. A reader should be able to quote them by number by the time they finish the book. The endpapers carry them, and every later chapter refers back. The principles are the spine; the practice is the body that hangs off the spine.

5. What this chapter asks you to give up

Every methodology worth adopting asks the reader to give something up. Most methodology books avoid naming the concessions, because naming them makes adoption harder. I'd rather name them.

If you adopt the inversion, you have to give up the idea that engineering skill lives primarily in the act of typing. This is the hardest one for most career engineers, because typing is what the industry trained them to be good at, and the industry rewarded them for being good at it for twenty or thirty years. The skill hasn't become worthless. But it has stopped being the center of the work. The center of the work is now specification and verification. A senior engineer who cannot write a tight acceptance contract is a senior engineer whose senior-ness is at risk. A mid-level engineer who can write tight acceptance contracts and curate sharp context bundles is about to have a different career trajectory than the one that was in front of them a year ago.

You have to give up the backlog-as-inventory habit. A backlog full of tickets is a work-in-process inventory, which in any other manufacturing context would be understood as a cost. Under CID, the pod carries a small number of contracted VOSes and a small number of in-flight ones. That's it. There is no backlog stretching out into next quarter. There is no Jira board with a hundred tickets the team will, in aggregate, never look at. If a VOS isn't contracted, it isn't real work yet. If it's contracted and hasn't shipped, it is the current work. The inventory habit comes from a world where typing capacity was fixed and speculative work had to be stockpiled against future capacity. That world is over. Keeping the stockpile is keeping a habit without the reason for it.

You have to give up the performance of sprint velocity. Velocity is a story the team tells itself about activity, not an outcome. A pod running CID at full capacity has no velocity number, because the unit of work (the VOS) is not interchangeable with other VOSes the way story points are supposed to be interchangeable. What the pod tracks instead is a shipped count, a first-pass verification rate, and an outcome-hypothesis-hit rate. Those numbers tell whether the pod is producing user-observable value and whether its predictions about that value are calibrated. That's the story an executive actually needs, and the velocity dashboard was trying and failing to tell it.

You have to give up the idea that engineering authority lives in implementation details. The HOW section is short on purpose, the WHAT constrains, the acceptance contract binds, and none of those documents tell the generation layer how to write a function or what to name a variable. The authority that used to live in those choices has moved. It now lives in the intent, the context, and the contract. An engineer whose professional identity is built on the variable-naming choices is an engineer whose professional identity is under pressure.

You have to give up some of the rooms. Standups, sprint reviews, PI planning events for thirty. Those were the scaffolding of a typing-bound process. CID doesn't need them. The pod at Alchemaize has not held a standup in fourteen weeks, not because we forgot, but because the coordination the standup was for, keeping humans whose typing had to stay synchronized in the same place at the same time, isn't the coordination the pod actually needs. What the pod needs is a verification gate, a context bundle, and a shared view of the intent stream. Those live in the repo.

And you have to give up a kind of comfort. The comfort of knowing what the job is, what skills to develop, why you're senior, all because those things have been stable for twenty years. The methodology isn't mean-spirited about that. It's trying to be right about what the work has actually become, and being right is uncomfortable for anyone midway through the old arc. I was midway through it myself, two years ago. I know what it cost to look at the new shape and accept it. I'm not asking the reader to do something I haven't done.

A career is a set of investments. The chapter is saying that some of those investments have depreciated and others, the ones in specification discipline, verification rigor, context curation, and outcome thinking, have just appreciated. Those skills were always present in good engineers. The methodology makes them the main thing. The credit side of the ledger is real, and the rest of the book is about it.

6. The handoff

The pipeline is drawn, the VOS has a shape, the lifecycle has states, the principles are numbered. Chapter 5, which Casey writes, goes deep on context curation, because context curation is where the quality of a generation separates a five-times productivity multiplier from a thirty-times one. Chapter 6, Glenn's chapter, goes deep on verification, because verification is the single architectural decision that prevents AI-speed generation from turning into AI-speed disaster. Chapter 7 puts the pipeline in motion in a pod, day by day. Chapter 8 is Casey's showcase, where the methodology is not explained but enacted.

The vocabulary and the images are now on the page. What remains is the practice.

Writing a WHAT in Gherkin when you've never written a test in your life

The first WHAT I wrote took me four hours.

I'd like to say it was four hours of careful thought. It wasn't. It was four hours of staring at the Gherkin examples David and Glenn had sent me, writing something, deleting it, writing something else, deleting that, and finally producing a document that was, as Glenn told me the next morning, not where it needed to be yet.

Glenn and David reviewed the acceptance contracts in that first VOS. A good chunk of them came back with feedback. On one of them I felt my perspective was stronger, we talked it through for about twenty minutes, and they came around to my version. On the rest, their feedback made the contracts sharper and I improved them based on what they'd flagged.

Here is what I learned from the contracts that came back with the most feedback.

The first one was a scenario that described what the software should do when a user clicks a button. I'd written: *Given the user is logged in, When the user clicks submit, Then the form is processed.* The feedback was that "the form is processed" doesn't describe an observable outcome. It describes an implementation event. A user doesn't see a form being processed. A user sees the next screen, or a confirmation message, or an error. "Processed" was a word I used because in a normal requirements document it would have been fine, and here it wasn't.

The second one was a scenario that had three scenarios pretending to be one. I had Given and When and then five different Then clauses, each describing a different outcome in a different part of the UI. The feedback was that if any one of those Thens failed, I wouldn't know which one. The scenario wasn't testable. It was a wish list of things that should all be true, not a falsifiable claim.

Both of those pieces of feedback taught me something I've come back to a hundred times since. Writing a WHAT in Gherkin isn't a testing discipline. It's a writing discipline. I'd done a lot of writing in my career. None of it had been this particular kind of writing.

Let me explain what kind.

The kind of writing where every clause is doing real work. No throwaway sentences, no setup phrases, no smoothing. Every Given sets up exactly one condition. Every When names exactly one action. Every Then names exactly one observable. If two things are true, you write two Thens. If the precondition has three parts, you write three Givens. Nothing compresses. Nothing gets merged into "the system is in a valid state." The system is in a valid state is not a Gherkin clause. That's an excuse for not writing the Gherkin clauses.

I'd been told for twenty years that what I was going to struggle with, if I ever tried to be closer to the engineering, was the technical stuff. The algorithms. The data structures. The architecture. None of that turned out to be the hard part. The hard part was giving up the sentence-softening habits I'd spent twenty years perfecting in a consulting job. The habits that let you send an email to a stakeholder saying *we really need to be careful about accuracy here* when what you meant was *if the math is off by a penny this is a critical bug and we don't ship.* The first version is what you send. The second version is what the software actually has to enforce. For twenty years my job had been to translate the second version into the first one. The Gherkin is asking me to do the opposite.

I'd been trained to write things that would not look rude in an email. The Gherkin asks for things that would look rude in an email.

That took me about two weeks to internalize. The first week I kept softening. The pod kept giving me feedback. The second week I started writing the way the document actually needed me to write, and the feedback started getting lighter.

Here's a concrete example of the shift. In week one I might have written: *Given the user has completed onboarding, When they upload a document, Then the document is validated.* The feedback would come back: "Validated against what." So I'd add: *Then the document is validated against the required fields.* More feedback: "Which required fields? For what type of document?" *Then the document is validated against the invoice rules defined in the onboarding flow.* Still more: "What does validated mean here. Does the document become accepted, does it get a flag, does it get rejected, does the user see something." By the time we'd worked through it together, the sentence read: *Then the document appears in the user's document list with a green checkmark if every required field is present, or with a red flag and a list of missing fields if any required field is absent.* That is the sentence the software can be verified against. The earlier versions were not sentences. They were invitations for an engineer to guess.

A few things I figured out along the way. I'll share them because I think they're useful for anyone starting out.

The happy path is not the contract. If all you write is the happy path, you've specified about a quarter of what the software has to do. The software will be generated to pass your happy path and will quietly fail on every case you didn't name. You have to name the errors. You have to name the edges. You have to name the boundaries. Every VOS I ship has at least three scenarios: the happy path, at least one error path, and at least one boundary case. Most have four or five. A VOS with one scenario is an unspecified VOS.

Numbers matter. If the contract says "a lot of," the contract isn't a contract. "A lot of users" is not Gherkin. "At least ten thousand users" is. Pick the number. If you don't know the number, the VOS isn't ready.

The word "correctly" is banned. If I write "the invoice is calculated correctly," that's not a scenario. Calculated against what. Correct by whose definition. Every time I catch myself writing "correctly" in a Then clause, I delete the whole sentence and rewrite it with the actual condition the software has to satisfy.

The Given is not optional. I used to skip the Given when the precondition seemed obvious. Glenn and David both caught this. What seems obvious to me is not obvious to the generation layer, and in any case, the generation layer isn't the only reader. Three months from now, another

Intent Engineer is going to read this VOS and needs the Given to understand what state the scenario starts in.

I've been told by a couple of people who've watched me work that the Gherkin discipline is a testing skill I somehow picked up fast. It isn't. It's a sentence-writing skill. I'd been writing sentences in operations memos for twenty years. What I'd been doing in those memos, without calling it by this name, was avoiding the Gherkin discipline on purpose. An ops memo that reads like Gherkin would get me in trouble with stakeholders. An acceptance contract that doesn't read like Gherkin will get me in trouble with the pod. I've learned to switch between the two registers. It turns out the switch is the skill.

Here's the part where David said I was allowed to disagree with him if he'd framed any of this as specification-driven development in an academic tone. I haven't read the earlier drafts of the chapter carefully enough to know if he did. But if the chapter used the phrase "specification-driven development" anywhere, I'd push back on it. What I'm doing isn't specification-driven development. That phrase makes it sound like a computer-science seminar. What I'm doing is writing down what I actually want, in sentences that would be embarrassing to show a lawyer because they're so specific they leave no room for interpretation. That's the whole job. It isn't a framework. It's a register of writing.

If you're coming to this from where I came from, consulting, operations, customer-facing work where you've spent years softening sentences so they land, the good news is you've already got the sentence-craft. You just have to flip it. Write the thing you usually avoid writing. The one your stakeholder would push back on because it's so direct. That sentence is the one the machine needs. It's not the tone you want in most of your job. It's the tone the acceptance contract calls for.

One last thing. The first time one of my VOSes went through the review on the first pass, without a single piece of feedback on the acceptance contracts, I caught myself waiting for the usual morning follow-up. It didn't come. I checked the folder. The VOS had moved to shipped overnight. I'd gotten it on the first try.

That felt really good. I smiled, took a breath, and started writing the next one. That's the rhythm once you find it.

I'm grateful David gave me the space to share this part of the chapter. Now let me hand it back over to him to continue.

CHAPTER 5: CONTEXT CURATION

1. The wrong instinct

Before I walk you through the methodology's rules around context curation, what to include and what to exclude and why it matters, I want you to know something upfront. Everything in this chapter can be done without reading a single line of code. I built several production applications over the hundred days this book covers, and I have never once opened a file and read through the code to understand what it does. I think in features, not files. I use AI to handle the technical resolution. Later in this chapter, in section 5, I'll walk you through exactly how I do that. But first, let me explain the methodology piece, because the rules of inclusion and exclusion matter regardless of whether you're an engineer who reads code or someone like me who doesn't.

The first thing almost everyone does, when they try CID for the first time, is wrong. They point the AI at the whole codebase. They figure more information is better. They figure the machine should read everything and pick what matters, because that's what a new hire would do if you gave them two weeks of reading time before they touched a ticket.

I did this too. My second week of actually writing VOSes, back in January, I included something like twenty files in a context bundle on The Trade Codex. I can look back at the commit and count them. Twenty. The AI produced code that ran. It did roughly what I'd asked. And it matched none of the patterns already in the application.

The tests I ran passed. The code looked fine on its own. Then Glenn and David reviewed it and flagged that the code was doing its own thing instead of following what we'd already established.

That's the thing about giving the AI the whole codebase. It doesn't read the codebase. It doesn't really read anything. It samples. It pattern-matches. If you hand it twenty files, it takes an average of those twenty files and produces something that averages with them. If two of the files use one convention and eighteen use another, you'd think you'd get the eighteen. You don't always get the eighteen. You get whatever the model decided was salient, and salience is not a count.

I want to say that differently because it matters. Larger context windows don't mean better outputs. They mean outputs that are optimized for more signal, most of which is irrelevant to what you're actually trying to do. If you're specifying a password-reset feature and you've loaded the image-upload module into the context, the model has now seen that you do image uploads in a particular way, and some of that will bleed. Not most of it. Some of it. Enough of it that the code you get back will have weird echoes of code you didn't ask about.

This is the part that took me a while to internalize, because my prior instinct was the opposite. I'd spent twenty years in enterprise software-delivery rooms. The working assumption there, when you brought a new engineer onto a project, was the more context the better. Hand them the architecture doc. Hand them the runbooks. Walk them through the deployment pipeline. Give them two weeks of reading. If they didn't ask enough questions, you worried about them. If they asked a lot of questions, you were reassured.

That instinct is exactly wrong for the AI. The AI is not a new engineer. A new engineer who reads the whole codebase is forming a mental model. The AI reading the whole codebase is doing something else. It's averaging. It's a different kind of reader, and the thing it rewards is different.

Matthew Skelton and Manuel Pais's *Team Topologies* makes the point that cognitive load is a team-design variable, something you budget for at the org level, not a personal failing to be shamed out of an individual engineer [1]. That frame helped me stop blaming myself for the bad bundles. The pod is the thing that carries the load. The bundle is how the pod shapes what the machine has to hold in its head on any given run. If the bundle is too big, the load is wrong, and the output is wrong, and neither fact is about the person drafting it.

The right frame, once I stopped fighting it, was this. Every file you put in the bundle is a vote. The model counts the votes. It doesn't weight them by how relevant you think they are. It reads what you give it and produces something that fits the neighborhood of what you gave it. So the bundle is not *the universe the AI needs to know about.* The bundle is *the neighborhood you want the output to match.* If you put a bunch of houses from the wrong neighborhood in there, the output looks like a hybrid.

The fix is to stop giving it everything. The fix is to give it the specific five to fifteen files that show the patterns you want the output to follow, and nothing else. Not the architecture doc. Not the runbook. Not the ninety other files in the module. Five to fifteen.

The first time I tried a small bundle, I cut my Trade Codex bundle from twenty files to six. The verification rate jumped. Not because the AI got smarter. Because the AI stopped being asked to average across stuff that didn't matter.

That was the moment the instinct flipped for me. Smaller bundles produce better code. The discipline is exclusion, not inclusion.

2. *What a good bundle looks like*

Let me walk through one. I want to pick a real one, not a sanitized one, because the sanitized version always looks obvious in retrospect and misses the part that's actually interesting.

The bundle is from The Trade Codex. The feature was the AI Trade Analysis, VOS #3 in the project. This was the feature that gives every completed trade intelligent feedback through Bedrock: pattern recognition, risk assessment, improvement suggestions. Four areas of context. Here's the list, and I want to be upfront about something before I walk through it: I used my AI tool to help me identify the specific files to include. I knew what I needed conceptually, but the tool helped me find exactly where those things lived in the codebase. That's part of how this works.

The database layer. Reason: the AI analysis feature has to read trade data to analyze it. I told the tool to include the files that handle how trades are stored and queried. If the AI doesn't see how trades are structured, it'll invent a data shape for trades, and its shape won't match ours.

The Bedrock AI client. Reason: we'd already built a client for talking to Bedrock's Converse API for other AI features. The new analysis feature has to use that same client, not create a new one. If the AI doesn't see it, it'll write its own Bedrock integration from scratch, and now we have two ways of talking to the same service.

The trade detail page. Reason: this is the page where the analysis result shows up for the user. The feature has to add an analysis section to this page. If the AI doesn't see it, it'll generate a new page or a wrapper component, and the wrapper won't match the layout patterns we've already established.

The analysis display component. Reason: we already had a placeholder component for where the analysis would render. The feature needs to fill this component with real content. If the AI doesn't see it, it'll create something new instead of building on what's already there, and the styling will be off.

That's the bundle. Four areas. Four reasons.

I want to pause here and say something about the language I just used. I said, "database layer" and "Bedrock AI client" and "component" like those are words I've always had. They're not. Before January, I would have said "the part that stores the trades" and "the AI thing" and "the screen where it shows up." Over the hundred days of working inside this methodology, watching the AI build things, reading what it produced, asking questions when I didn't understand something, I started picking up the vocabulary. Not because someone taught me a curriculum. Because I was in the room every day watching the patterns, and the patterns have names, and after a while you start using the names because they're more precise than the descriptions you were using before.

That's important context for the reader, because if you're coming to this chapter without an engineering background, the way I talk about files and modules and components now is not the way I talked about them in January. The methodology teaches you the language as a side effect of doing the work. I didn't set out to learn what an API is. I set out to build a trade journal, and the API was part of what got built, and now I know what it is because I watched it get created and I asked what it was doing.

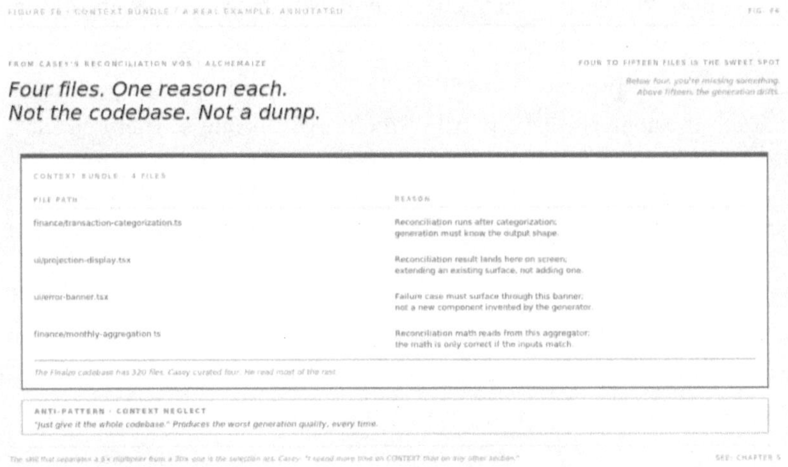

Figure F6 · CONTEXT bundle example. *Annotated example*

A few things to notice about those reasons. None of them is *this file might be relevant.* None of them is *I touched this file last week.* Every one of them is a sentence about what the generated code needs from that area of the codebase. The data shape it needs to match. The client it needs to reuse. The page it needs to slot into. The component it needs to fill.

This is the whole discipline. Every area in the bundle answers the question *what would go wrong if I didn't include this?* If I can't answer that question in plain language, it doesn't go in.

I want to say what's *not* in the bundle, too, because that's actually where the skill is.

The Trade Codex codebase, at that point, had well over a hundred files. Of the ones that weren't in the bundle, a lot were things I might have been tempted to include a few weeks earlier. The entire authentication system, for instance. It's reasonable to think the AI should know how login and user sessions work. But the AI analysis feature doesn't touch authentication. If the AI sees the auth system, it might start thinking about user identity in an analysis context that doesn't care about it. So auth stays out.

Same for the database migration files. Same for the deployment configuration. Same for the Seer engine, which is a separate automated strategy system that matters for other features but not for this one. Same for the campaign and quest modules. Same for the other display components that aren't the analysis display. All of them exist. None of them go in.

The bundle is not *everything relevant to The Trade Codex.* The bundle is *everything this specific slice of code needs to consult.* Those are different sets, and the second one is much smaller than the first.

I remember a point in week three or four where I realized I was spending more time deciding what to leave out of bundles than what to put in. That was a good sign. The skill is the exclusion.

Here's one more thing, and it's a small thing but it matters. I write the reasons next to the files, in the VOS document, in a line each. This is not for me. It's for Glenn, who reads the VOS during verification. If he sees a file with no reason, he'll ask why it's there. And if the reason is weak, he'll push back, and I have to defend it or take it out.

The reasons force me to be honest. "Because I thought it might help" is not a reason. "Because it defines the data shape the new code has to match" is a reason. The first time Glenn asked me why a file was in a bundle and I couldn't answer, I took it out. The answer turned out to be *no reason*, which meant it shouldn't have been there, which meant I'd been in the old habit.

Every file in the bundle has a stated reason. That's the rule.

3. The skill ladder

When I started in January, my first-pass verification rate was around 40 percent. That means, of every ten VOSes I wrote, four produced code that came through review clean on the first generation and six came back with feedback. The six that came back did so for various reasons. Sometimes the acceptance contract was underspecified. Sometimes the context bundle was off. Sometimes both.

I remember thinking 40 percent wasn't terrible. I'd expected worse, because I was learning the form. And I was learning two things at once, the acceptance contract discipline and the context curation discipline, and it was hard to tell in week one which of the two was costing me the failures.

By week three I could separate them. I could look at a VOS that came back with feedback and know whether the Gherkin had been the issue or the bundle had been the issue. In my case, the bundle was the more common cause. The Gherkin I was actually pretty good at, because I'd written enough requirements documents in my career that saying precise things in structured form was not a new skill for me. The bundle was the new skill. I'd never had to decide, before, which files a piece of software should consult. Engineers had done that implicitly by sitting at a computer and reading what they knew to read. I'd been in the room for that, but I'd never done it.

Week 1 through 2, my bundles were too big. I was including ten, twelve, fifteen files when six would have done. The extra files were doing real damage. I wasn't seeing it yet. Verification rate in the 30 to 50 percent band.

Week 3 through 4, I started cutting. I'd draft a bundle at twelve, then re-read it with a question in mind, and cut it to eight. Sometimes to six. The reason-per-file discipline emerged here. I'd write the reasons down and realize three of them were weak, and I'd cut those three. Verification rate in the 50 to 70 percent band.

Week 5 forward, something shifted. I stopped drafting bundles by addition and started drafting them by question. The question was *what does this slice of code need to consult to be correct?* I'd ask the question, and the answer was usually four files. Sometimes three. Sometimes six. Rarely more. I'd write those files down, write the reasons, and the bundle was done in a few minutes. Verification rate in the 70 to 90 percent band.

By week eight or nine, on my third or fourth app, I was consistently in the 80 to 95 percent range. Glenn and David stopped flagging context bundles as the main area for improvement, because they mostly weren't any more. When a VOS came back with feedback at that point, it was usually an acceptance contract issue or a genuine ambiguity I hadn't resolved.

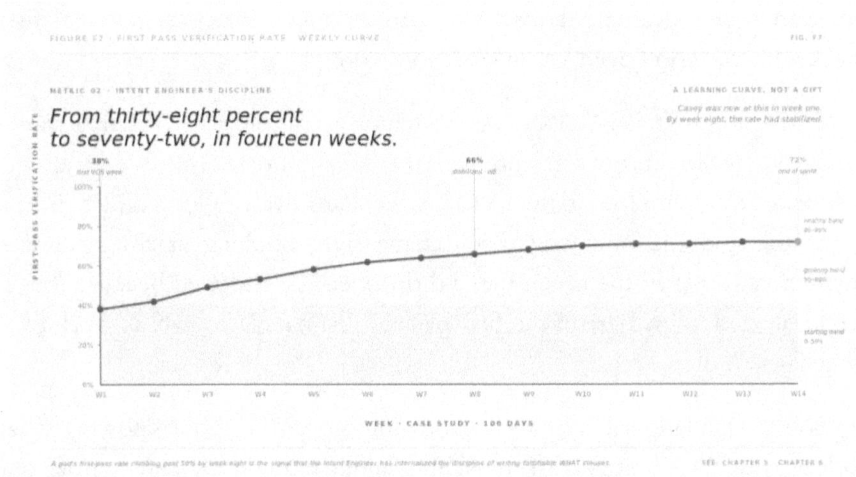

Figure F7 · First-pass verification rate curve. *Weekly rate climbing from 30-50% through 80-95% band.*

That's the ladder. I don't want to make it sound more precise than it was. The week numbers are rough. Some weeks were better than others. I had a stretch in February where my rate dipped because I was working on a new module I didn't know as well as the parts I'd been working on, and I hadn't done the reading yet. (I'll come back to that in section 5.) The curve isn't smooth. It's a trend.

But the trend was real, and the shape of it surprised me. I thought I'd get better at bundles by learning to include more relevant files. What actually happened was I got better by learning to exclude more irrelevant ones. Week 1, I had twelve files and was trying to figure out which one I was missing. Week 5, I had four files and was asking whether any of them could come out.

The skill is exclusion. I want to say that again because it was the thing I kept not believing until I watched my own rate move.

One thing that helped me see the trend was keeping track of it. I didn't build a spreadsheet or anything. I kept a note in the VOS directory of how many files I'd included and whether the VOS came through review clean on the first try. After about thirty VOSes you can see the pattern. Fewer files correlated with cleaner first reviews, for me, once I was past the very first week where I didn't even know what I was doing.

One thing that also helped was watching what happened when I re-did an old VOS. Sometimes a feature would need an update a few weeks after it shipped. I'd pull up the original VOS, which had twelve files, and I'd draft a new one, and the new one would have five. Looking at that gap was instructive. Most of the seven files I'd dropped, I'd dropped because they'd never needed to be there in the first place. I just hadn't known, in week two, which seven they were.

So the skill ladder is not about knowing more. It's about knowing what to leave out. The first part of the skill is understanding the features of your application well enough to identify what areas matter for a given VOS. The second part is the willingness to leave the rest out. The second part is harder than it sounds, because leaving things out feels like risk. Every file you exclude is a file the AI can't consult. The instinct is to hedge. The discipline is to not hedge.

Glenn and David both said something to me around week four, after I'd been pushing bundles from twelve files down toward seven. Glenn said something like, *you're going to be better at this than David or I are, because you don't have the engineer's instinct to over-include as protection against the unknowns you can see.* I think he's partly there. I don't have the engineer's over-include instinct, but I also didn't have the judgment about what the application needed in the early weeks. I had to build that. The skill is the exclusion plus the judgment about what to exclude. Neither alone is enough.

I'll say one more thing about the ladder and then move on. The numbers I've given (40 percent in week one, 80 to 95 percent by week five onward) are specific to me, and they're specific to the apps I was on, and they're specific to working with Glenn as the verifier. Your numbers will be different. The shape of the curve is what generalizes. You'll start in the 30 to 50 range, you'll move to 50 to 70 over a few weeks, and you'll sit somewhere in the 70 to 90 range once the skill is formed. Beyond that, a person doing this work seriously for a quarter or more lands in the 80 to 95 range. Where you personally settle depends on the codebase, the verifier, and how much reading you're willing to do.

If your rate isn't moving, something's off. Either you're not writing down reasons for each file, or you're not cutting the files whose reasons are weak, or you're working on a codebase you don't understand well enough yet. Usually it's the third one, which is section 5.

4. The stated-reason discipline

This is what we created as a rule for this methodology. I'm going to walk through it and then explain why it matters.

Every file in a context bundle has a stated reason for inclusion. The reason is written into the VOS document, next to the file reference, as a short sentence. "Just in case" is not a reason. "Because I touched it last week and it might be relevant" is not a reason. "Because it defines the data structure the new code must preserve" is a reason.

That's the discipline we landed on. I've found no exception to it in a hundred-plus VOSes.

Here's why it works, from my experience. When I started writing reasons, I thought they were bureaucracy. A paperwork item Glenn wanted because he's rigorous about verification and I was the new one. I wrote the reasons because I was asked to, and I assumed the reasons were for Glenn. They weren't for Glenn. They were for me.

What the reasons did, once I started writing them, was force me to ask a question I hadn't been asking. Why is this file here? I'd put twelve files in the bundle by instinct, and when I went to write the reasons, I'd get to reason number nine and realize I didn't have one. The reason I had was *because I touched this last Tuesday*. That's not a reason the generated code cares about. That's a reason I cared about. The file came out.

Reason number ten was *because the code review last week referenced it.* Out. That's a reason a human might want that file; the machine doesn't benefit from the reference.

Reason number eleven was *because it's in the same directory as the file I'm modifying.* That's shared-neighborhood thinking. Out, unless the file being modified depends on that file directly.

By the time I was done writing reasons, four or five files were left. Those were the ones with honest reasons. The bundle was done.

I want to say what a good reason looks like, concretely, because *write a reason* is vague advice. Here are reasons I've written that held up:

Because the new code has to produce output in this shape and this file defines the shape.

Because this is the only existing implementation of the pattern the AI needs to follow, and without it the AI will invent a different pattern.

Because the tests in this file show the testing convention the new tests have to match.

Because the error state we're going to surface has to use this component, not a custom one.

Those are the shapes. What's true of all of them is that they name the thing that would go wrong if the file weren't in the bundle. *If this file isn't here, the code will be shaped wrong, styled wrong, tested wrong, or won't match the existing pattern.* That's the test.

Here are reasons I've written that didn't hold up:

Because it might be relevant.

Because I've been working in this area.

Because Glenn suggested I look at it once.

Because it's the main file in the module.

Because the file name sounded related.

None of those survive contact with the discipline. They're associations, not consultations. They're reasons I had in my head. They're not reasons the generated code benefits from.

Writing the reasons down makes you confront your own uncertainty. The AI can't carry your uncertainty for you. The AI will just read what you gave it and produce something consistent with it. If the thing you gave it is a confused average of relevant and irrelevant files, you get confused-average output. That's not the AI's fault. That's the bundle.

I'll say the dry version because it's true. If you can't finish the sentence "this file is in the bundle because __________" in fewer than fifteen words, the file doesn't belong. If the sentence you finish is "because I have a vague feeling," the file doesn't belong. If the sentence starts with "because last week," the file doesn't belong. The file belongs when the sentence is about what the generated code has to do, and how the file's presence changes what the generated code does.

Glenn once described this as applying the same discipline to inputs that we apply to outputs. I like that framing. The acceptance contract shapes what we expect coming out of the generation layer. The stated-reason discipline shapes what we put in. Both exist for the same reason. Clarity at the edge. What the machine works on is only as precise as we made it. The machine doesn't add precision. It runs on what we gave it.

5. What to do when you don't know what to include

There's a real problem I want to address, because everything I've said in this chapter assumes you have a sense of what features and functions exist in your application. A lot of the time, especially when you're starting out or moving to a new project, you don't. I certainly didn't when I started building out the deeper features of The Trade Codex, and I had to figure out what to do about it.

I want to be honest about something here, because I think it matters for the reader who is coming to this without an engineering background. I have never read a codebase. Not once. I have built several applications over the hundred days this book covers, and I have not once opened a file and read through the code to understand what it does. That is not how I work. I use an AI agent for that. When I need to understand what exists in the application, I ask the AI to tell me. When I need to know what features are connected to what, I ask. When something isn't working, I don't go troubleshoot the code myself. I describe what I'm seeing and let the AI agent figure out where the issue is.

So when I say "understand what's in your application before you write a VOS," I don't mean go read the code. I mean understand the features. Understand what the application does at a high level. Know what the major pieces are and roughly how they connect. You can get that understanding by using the application, by talking to your AI agent about it, by looking at the feature list, by reviewing the VOSes that have already been shipped against it. You don't have to read a single line of code to build that understanding.

Here's what I actually did when I started working on the deeper features of The Trade Codex in mid-February. The app had grown past the initial trade journal into AI analysis, campaigns, quests, and market data. I needed to understand what was already there before I could write good VOSes for new features.

I spent time with the application itself. I used it. I clicked through the screens. I looked at what was there and what wasn't. I asked my AI agent questions like "what features do we have for trade analysis right now?" and "how does the campaign system connect to the trade journal?" I built a mental model of the application by interacting with it and asking questions about it, not by reading code.

When it came time to write a VOS for a new feature, I'd tell the AI agent what I wanted to build and what existing features it needed to work with. The AI agent would help me identify which files and modules were relevant. I'd describe the context in human terms ("include the files related to how we store trades, include the AI analysis client we already built, include the page where this is going to show up"), and the tool would resolve that to the actual files.

The discipline I developed was not about knowing the codebase. It was about knowing the features. If I understood that The Trade Codex had a trade journal, an AI analysis system, a campaign system, a quest system, a Seer engine, and a theme system, I could scope my context bundles by feature area. "This VOS is about adding a new type of AI feedback on trades, so I need the files related to AI analysis and the trade detail page. I don't need the campaign files. I don't need the Seer files. I don't need the theme files." That's feature-level thinking, and it's the skill that actually matters for context curation.

The AI tool does the translation from "files related to AI analysis" to the actual file paths. That's what the tool is good at. What the tool is not good at is knowing which features are relevant to your VOS. That's the human judgment. That's the Intent Engineer's contribution to the context bundle.

I want to be clear about this because I think the instinct for a lot of readers, especially those with engineering backgrounds, will be to tell people to go read the codebase first. That advice made sense when humans were writing all the code and needed to understand every line. In the methodology we're describing, the AI handles the code. What the human needs to understand is the product: what it does, what the pieces are, how they connect at the feature level. That understanding comes from using the application and asking good questions, not from reading files.

If your context bundles aren't working and your VOSes keep coming back with feedback, the fix is usually not "go read more code." The fix is usually "understand the feature landscape better." Ask your AI agent to walk you through what exists. Use the application yourself. Review the shipped VOSes. Build the mental model at the feature level and let the AI handle the file-level resolution.

That said, the educational content in this chapter about what makes a good bundle still applies. The stated reasons, the exclusion discipline, the five-to-fifteen file range, all of that travels. The difference is in how you get there. An engineer might get there by reading the codebase and knowing which files matter. I get there by understanding the features and letting the AI agent identify the files. Both paths lead to the same bundle. The methodology works either way, and that's part of the point.

6. The takeaway

If you do one thing differently after reading this chapter, try this. Cut your next context bundle in half, write a sentence for every file that remains, and if you can't write an honest sentence for one of them, take it out.

Then track your first-pass verification rate for the next two weeks, because the rate will tell you whether the discipline is forming. If the rate is going up, keep doing what you're doing. If it isn't, the reasons are probably too soft, and the thing to do is push harder on the reasons.

That's the practice. Five to fifteen files, every one with a stated reason, and the willingness to leave the rest of the codebase out. That's what a good bundle looks like, and that's what the skill eventually becomes, and I can tell you from the other side of it that it is a very different kind of work than the industry has been asking operations and specification people like me to do.

Curating context is the pod's job, not the backlog's. *Team Topologies* frames the stream-aligned team as the unit that owns its own cognitive boundary [1]; in CID, the bundle is where that boundary is drawn every time a VOS runs. Nobody upstream of the pod can draw it for you. Nobody downstream can fix it after the fact. It is the pod's work, done at the start of the turn, and the quality of the turn rides on it. I know that because I watched the quality of my own turns start riding on it around week three, and I've been watching it do the same thing for other Intent Engineers on other pods ever since.

Notes

1. Matthew Skelton and Manuel Pais. *Team Topologies: Organizing Business and Technology Teams for Fast Flow*. IT Revolution, 2019.

CHAPTER 6: VERIFICATION UPSTREAM

1. The reckoning at week three

I have watched this happen more times than I want to count over the last fourteen months. The results are always the same.

A team adopts an AI coding tool. For the first two weeks, everyone is happy. Velocity doubles, then triples. The backlog empties faster than it fills. Engineers who were burning out on ticket triage get to claw back a week and use it to clean up code they had hated for a year. The product people are delighted. The CFO starts asking whether the next hire is still necessary.

Then week three arrives, and the bug count goes vertical.

Not a few more bugs. Not a concerning drift. Vertical. A team that was shipping ten features a sprint is now shipping fifteen features a sprint *and* carrying forty new defects into next sprint instead of eight. That's fifty percent more output. Five times the bugs. The defects are subtle. The obvious ones got caught. The ones that leaked through are the kind that look OK in a PR diff only to fail once a real user trips them, which the team finds out two weeks later when the customer support queue catches fire.

The engineering lead usually responds the way they've been trained to respond. They add more review. More eyes on each PR. More tests before merge. More staging environments. Maybe even a new QA hire if there's budget for it.

It makes sense. That's the playbook that worked for years. Slow things down a little, put more humans in front of the code, and catch problems before they ship. But that playbook starts to break in an AI-assisted workflow.

The issue isn't that people suddenly forgot how to review code. The issue is that the volume changed. A QA process that could keep up with one developer writing 100 lines a day was never designed for that same developer reviewing, accepting, and merging 2,000 lines of AI-generated code in an afternoon.

The review lane is still the same size. The development lane just got dramatically wider.

And you can't solve that by simply hiring more people. Not for long, anyway.

That's the week-three reckoning. At first, it looks like a quality problem. But when you lift the hood, it's really a plumbing problem.

Downstream QA made sense when code was expensive to produce.

Back then, code moved through the system slowly. A developer wrote a limited amount of it, another person reviewed it, QA tested it, and the whole process mostly held together. Testing was a small tax on something scarce. It added time, but it also caught enough problems to justify the cost.

AI fundamentally changes that balance.

Now code is cheap to generate. Very cheap. A developer can produce more code in an afternoon than the old review and QA process was designed to handle in a week. So the old tax is still being charged, but it is being charged against a completely different kind of system. The cheap part of the workflow is moving faster and faster, while the expensive part is still moving at human speed.

That is why downstream QA starts to break.

The answer is not just to make QA faster. QA was already the narrow part of the pipeline. Asking it to keep up with AI-generated code by doing the same thing harder is not a reasonable strategy.

The better answer is to move verification earlier.

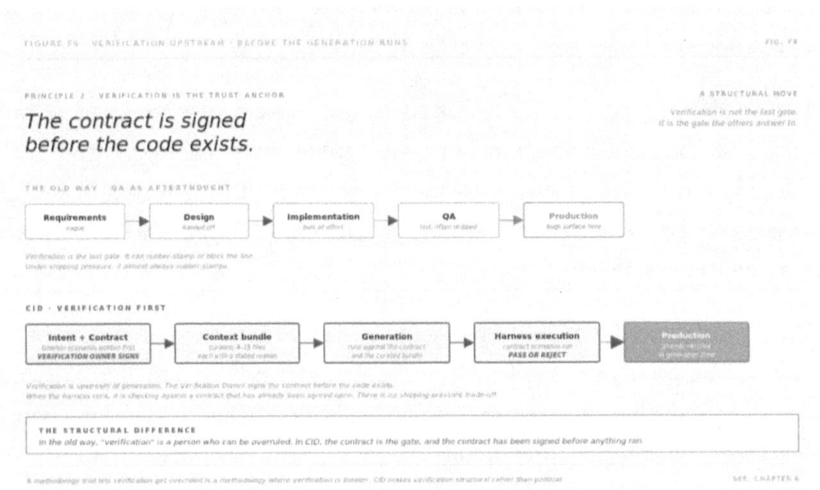

Figure F8. *Verification Upstream*

Before the code is written, the system needs to know what "correct" means. While the code is being generated, it needs to be checked against that definition. By the time the code reaches QA, the most important questions should already have been answered.

QA still matters. But it can no longer be the first place where correctness is seriously tested.

Here is the shift: In an AI-assisted workflow, verification must start before the first line of code is generated.

This is what I mean by moving verification upstream.

Not making downstream QA faster.

Not asking reviewers to work harder.

Not adding one more stage to an already crowded pipeline.

Moving verification upstream.

That reframe matters, because the rest of this chapter builds on it.

Before we move on, though, there is one more thing to say about week three. The team that runs into this problem is not being careless. They are usually doing exactly what they were taught to do.

For the last decade, most of our software delivery playbooks were built around one basic assumption: human typing was the expensive part. Developers wrote code slowly. Reviewers reviewed it. QA tested it. Release processes wrapped controls around it. DevOps, Agile, and continuous delivery all made sense in that world.

But AI changed the cost of producing code without changing the systems around it. The code got cheaper. The pipeline did not get smaller. So when the team hits the week-three reckoning, it is not because they are bad at engineering. It is because they inherited a workflow designed for a different reality.

The old pipeline assumed code generation was scarce and slow. AI made code generation fast and abundant.

That is the stale assumption.

A lot of my consulting work around CID has been about helping teams see that shift clearly. Once they see it, the next question becomes much easier to answer:

What should a pipeline look like now?

The rest of this chapter is about that new pipeline.

2. Acceptance contracts

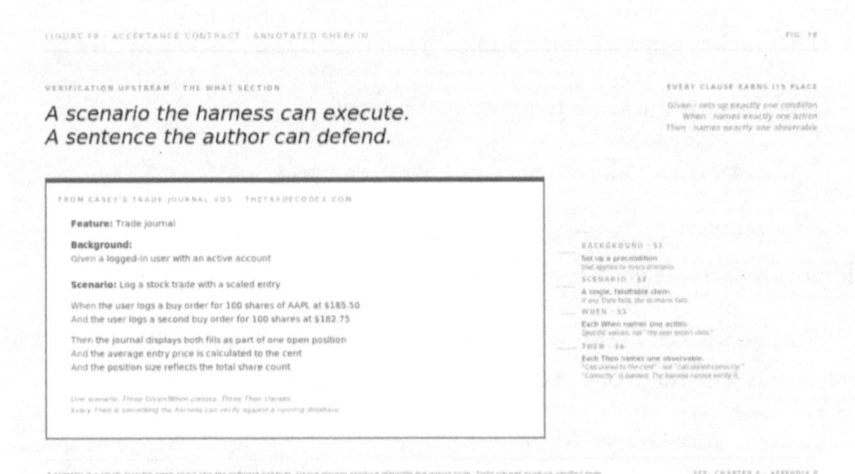

Figure F9. Acceptance Contract

An acceptance contract is the specification the machine can actually use.

It tells the system how a piece of software is supposed to behave *before* the code is generated. In a Verifiable Outcome Slice (VOS), it sits right after the WHY. The WHY explains what we are trying to accomplish. The acceptance contract explains what must be true for the work to count as correct.

That distinction matters.

An acceptance contract is not a test plan.

A test plan tells a human what to do. Click this button. Enter this value. Check this screen. Confirm this message appears.

That has value, but it is procedural. It depends on a person walking through the steps.

An acceptance contract is different. It describes the conditions the software must satisfy. It does not just say what someone should test. It says what must be true when the feature runs.

That makes it executable.

It is also not the same thing as a user story.

A user story explains intent from the user's point of view:

"As a billing administrator, I want invoices generated automatically so I do not have to create them by hand."

That sentence is useful. It belongs in the WHY. It tells us what the user needs and why the work matters.

But it is not enough to verify the software.

You cannot reliably turn that sentence into a pass or fail result. It is too human, too open-ended, and too dependent on interpretation.

An acceptance contract closes that gap. It turns the intent into specific conditions the system can check.

That is why we write acceptance contracts in Gherkin.

Given. When. Then.

Feature blocks. Scenario blocks. Background setup when several scenarios share the same starting point.

Gherkin was originally designed for Behavior-Driven Development, so non-engineers could help define acceptance criteria in a readable way. But it turns out to be useful for something else too: telling a language model exactly what to build.

That is the important part.

Gherkin is structured enough to limit the model's freedom. It makes the expected behavior clear before generation starts. But it is still readable enough that a product owner, domain expert, or Intent Engineer can write meaningful behavior rules without turning the whole thing into code.

That balance is what makes it useful.

It is human-readable, but machine-checkable.

It gives the model less room to guess.

And in an AI-assisted workflow, less guessing is the point.

Here is a real example. It comes from the HIPAA audit-trail rule in our compliance extension. I use this scenario when I teach the material because it is short enough to read quickly, but strict enough to show why acceptance contracts matter.

```
Feature: Patient Record Access Audit Trail
  Rule: HIPAA-04. Every access to Protected Health Information
  must be logged to an immutable audit trail sufficient to
  reconstruct who accessed what, when, and from where, per
  45 CFR §164.312(b).

  Background:
    Given the audit trail is configured to write to the
      append-only audit store
    And the clock is synchronized to a trusted time source
    And a patient record for "Jane Doe, MRN 0001" exists

  Scenario: Successful PHI access is logged
    Given a clinician "Dr. Aronoff" is authenticated with
      role "attending physician"
    And the clinician's session has a valid authorization
      for the patient's care team
    When the clinician retrieves the record for "Jane Doe, MRN 0001"
      from IP address "10.4.17.22"
```

```
  Then the response body contains the patient record
  And an audit entry is written within 250 milliseconds with the
    following fields populated:
      | field          | value                                 |
      | actor_id       | Dr. Aronoff's provider identifier     |
      | actor_role     | attending physician                   |
      | subject_mrn    | 0001                                  |
      | action         | READ                                  |
      | resource       | patient_record                        |
      | source_ip      | 10.4.17.22                            |
      | timestamp      | the request time, UTC, millisecond    |
      | outcome        | success                               |
      | correlation_id | matches the request's correlation id |
  And the audit entry is written to the append-only store
  And the audit entry cannot be modified or deleted by any
    application-level identity

Scenario: Denied PHI access is logged
  Given a clinician "Dr. Aronoff" is authenticated with
    role "attending physician"
  And the clinician's session does not have authorization
    for the patient's care team
  When the clinician retrieves the record for "Jane Doe, MRN 0001"
  Then the response status is 403
  And no patient record is returned in the response body
  And an audit entry is written with action "READ",
    outcome "denied", and the same nine fields populated
    as the success case

Scenario: Audit write failure blocks PHI disclosure
  Given the append-only audit store is unavailable
  When the clinician attempts to retrieve the record for
    "Jane Doe, MRN 0001"
  Then the response status is 503
```

```
And no patient record is returned in the response body
And the failure is recorded to the local degraded-mode
  buffer with intent to replay
```

Look at what the contract does, and just as importantly, what it does not do. It names the behavior that must be true. The audit entry must contain the required fields. The expected values must be present. The write must happen within the latency boundary. The audit store must be append-only. If the audit write fails, the protected action must be blocked. The record must be immutable, and the denied path must be covered, not just the successful one.

Those are not suggestions. They are predicates. Each one is something the generation stage must satisfy and the verification stage must check. But notice what the contract does not say.

It does not say how the audit entry should be serialized. It does not name the storage engine. It does not decide whether the replay buffer lives in memory or on disk. It does not tell the system how to generate the correlation identifier. It does not even dictate whether the timestamp is captured when the request arrives or when the audit record is written.

Those are implementation choices. The generation stage is allowed to make them, as long as the required behavior is preserved. That is the point of the acceptance contract. It does not tell the machine how to build the feature. It tells the machine what must be true when the feature is done.

This is where acceptance-contract work usually goes wrong the first time. New authors tend to over-specify the how and under-specify the what. They tell the machine which library to import. They describe the database table. They name the service. They give it implementation details.

But they forget to say the thing that actually matters: If the audit write fails, the disclosure must not happen. That is the whole point of the rule. And the machine will not automatically infer it. It may do exactly what it was told to do. It may import the right library, write to the right audit store, and produce code that looks correct at a glance. But if the code serves the patient record first and only then tries to write the audit entry, the system has already failed.

The disclosure happened. The log did not. That is the violation the rule was meant to prevent. So the contract has to say it plainly. The protected action must be blocked if the audit event cannot be recorded.

This is why the Verification Owner exists. Acceptance contracts are not just paperwork. They are the place where a requirement becomes something the system can actually check. In compliance work, they are the interface between the statute and the executable predicate. That translation is not trivial.

The Intent Engineer may draft the contract. But the Verification Owner reviews it and asks the harder question: Where is this contract silent in a way that would let a bug through? That is the work. Not making the contract longer for its own sake. Not turning it into a design document. But finding the missing constraints that matter.

Do that review enough times, and the team starts to build a body of acceptance contracts it can trust. That is what lets the pipeline move at AI speed without letting correctness become optional.

One more thing is worth clarifying. The HIPAA-04 example is a compliance acceptance contract. Most acceptance contracts are not. Most are feature contracts. The structure is the same. The source of the constraint is what changes. A feature contract gets its constraints from product intent. A compliance contract gets its constraints from regulation. A security contract gets its constraints from the threat model.

But they all serve the same purpose. They are written in Gherkin. They are executable. And they belong before generation, not after it. Chapter 11 goes deeper on compliance extensions. For now, the important point is simpler: The contract is the artifact. And the contract is upstream.

3. TDD as enforced protocol

Test-driven development has been around for decades. Almost every engineering team I have worked with says they do it. Most of them do not really do it. Or at least, they do not do it in the strict sense.

What usually happens is simpler: the developer writes the implementation first, then writes the test afterward. The code was faster to write. The test felt like overhead. So, the test becomes something added at the end to confirm the thing that already exists.

That is not ideal, but in a human workflow it is often forgiven. A good engineer still understands the behavior space. Even if they write the test after the code, they usually know what the feature is supposed to do, where the edge cases are, and what could go wrong. Their judgment fills in some of the discipline the process skipped.

The machine does not have that judgment. When the machine writes the test after the implementation, the test usually becomes a mirror of the code it just wrote. It does not really challenge the behavior. It confirms the artifact. That is the problem. Every assumption the machine made during implementation gets carried into the test. The test passes because it was written to pass. It may look like verification, but it is not actually catching much.

This is why TDD in CID is not just a guideline. It is a protocol. The machine is told, inside the VOS task plan, to write the failing test first. Then it must run the test and confirm that it fails for the expected reason. Only after that does it write the minimum implementation needed to make the test pass.

Red. Green. Refactor.

That cycle is not a style preference. It is the only version of the loop where the test has a real chance to act as an independent check on the implementation. When a human writes the red test first, their judgment usually produces a useful test. When the machine writes the red test first, the test captures what the acceptance contract said. Only that. And that turns out to be useful.

There is no folklore in the test. No old habits from a previous codebase. No hidden assumptions from the engineer who happened to write it. The test reflects the contract. If the contract is right, the test has a clean source of truth. If the contract is wrong, the test is wrong in the same way, and the mistake is easier to see.

There are two secondary rules enforced with this protocol. The first is that the red-green cycle should run only against the file, function, or slice under test. Not the full suite. Running the entire suite on every small TDD iteration turns the inner loop into a waiting room. It slows everything down and teaches the machine to batch work that should happen one step at a time.

The second rule is about mocks. Mocks are allowed for external services. They are not allowed for your own internal modules or your database. That line matters. A mock of an outside service is often practical. A mock of your own module can become a lie. It creates a seam that may not actually exist. And if the machine invents that seam just to make the mock work, the test is no longer testing the system. It is testing the machine's invention.

Use the real database. Use the real internal modules. If doing that is too slow, the problem is probably the test harness. The mock is hiding that problem, and hiding it is worse than fixing it.

This is usually where engineers push back. The objection sounds something like this: "I already do TDD. Why are we making such a production out of it?" The answer is simple. The machine is not you.

A senior engineer can recover from an ambiguous process because they have years of judgment to fall back on. The machine does not. If the protocol is not written down, the machine will invent the most plausible shortcut. And the shortcut is almost always whatever makes the current task finish in the fewest steps.

The machine will happily write the test last. If you let it.

I had one engagement where the pushback was strong enough that we ran a comparison. We built two similar features over the course of a week. Same team. Same model. Same acceptance contracts. One feature used enforced TDD. The other used the engineer's preferred flow: write the code first, then generate the test to match.

The enforced-TDD feature took about 15 percent longer to produce. But it caught three real defects during the red phase. The other feature finished first. Then it spent the second half of the week in rework after those defects surfaced in staging. We did not need to run the comparison again.

That team uses enforced TDD now.

Enforced TDD is the cost of admission for AI-speed generation. It is also what makes that speed safe to take.

4. *The Verification Owner*

The Verification Owner owns the acceptance-contract standard, the verification harness, the compliance extensions, and the rollback infrastructure. It is the most senior technical role on the pod, and it cannot sit inside the delivery chain.

That last part matters.

If the Verification Owner reports to the delivery lead, the role is compromised before the first hard decision arrives. Under schedule pressure, someone will eventually ask for the next VOS to move forward with a known gap. Sometimes the ask is direct. More often, it is quieter than that. A shrug. A raised eyebrow. A comment about the customer waiting. A reminder that the release date has already moved twice.

But the ask always shows up.

It shows up because the pressure inside delivery only moves in one direction. Delivery wants the work shipped. That is not a character flaw. That is the job. But when the person responsible for saying no reports to the person feeling the delivery pressure, the pressure usually wins. Not every time. Not every week. But often enough that, over a quarter, the signal starts to decay.

The structural fix is simple: take the reporting line out of delivery.

The Verification Owner can report to the CTO. Or to a Verification Officer at the enterprise layer. Or to a chief technology and risk function. Or to whatever equivalent function exists inside the organization. The exact title is not the important part. The important part is that the role does not report into delivery.

That is the structural requirement. Everything else is method and judgment.

The role has three core responsibilities.

First, the Verification Owner sets the acceptance-contract standard. That includes the contract template, the review rubric, the required-scenario checklist for each VOS type, and the Gherkin style rules that keep contracts readable as the pod matures and the backlog grows.

The Verification Owner does not write every contract. The Intent Engineer drafts the contract. The Verification Owner reviews it. That review rejects contracts that are incomplete, ambiguous, or not actually testable.

The rejection must be specific. Not "this needs work." That is not useful. The feedback should sound more like: this scenario is missing; this assertion cannot be verified; this field needs an expected value; this denied path is not covered.

Then the Intent Engineer revises and resubmits.

Early in a pod's life, three to five revision cycles is normal. That is not failure. That is calibration. Over time, the Intent Engineer starts to internalize the rubric. By month three, most contracts should pass on the first submission.

Second, the Verification Owner owns the verification harness and the rollback path.

The harness is the machinery that runs the acceptance contract against the generated code. It runs the static analyzers. It runs the security scanners. It runs whatever compliance extensions the VOS activated. Then it produces one clear signal: pass or fail.

That harness is not just a rented service. It is software the pod maintains. It has to be under the Verification Owner's control because every new contract type, every new compliance requirement, and every regulatory change can require the harness to evolve.

The rollback path is the other half of that responsibility.

A shipped VOS must be reversible at the artifact level, without manual edits, within minutes. If the Verification Owner cannot roll back a shipped VOS inside five minutes, they do not really have the confidence to approve it in the first place.

The rollback path earns the shipment decision.

Third, the Verification Owner signs off.

Every VOS moves to SHIPPED through the Verification Owner's signature. That signature is not ceremonial. It is a mechanical approval attached to the VOS record. It says the harness passed, the acceptance contract was met, the compliance extensions raised no blocking findings, and the rollback path was exercised during the release dry run.

That signature is the trust anchor.

Every later claim CID makes depends on it. The throughput numbers. The compliance posture. The metrics. The organization's confidence that AI-speed generation is not just fast, but controlled. All of it rests on the assumption that the Verification Owner's signature is a real signal.

It is real because the person producing it reports outside the delivery chain and has nothing to gain from signing off on a bad VOS.

This is why the role is a promotion from QA, not a lateral move.

The pay, seniority, and organizational placement must reflect that. If an organization fills the Verification Owner seat with a QA lead, keeps them at the old pay band, and leaves them in the old reporting line, it has not created the role. It has renamed one.

That renaming is one of the most common failure modes in CID adoption. Chapter 12 covers it directly.

So who is qualified to hold the role?

Someone with ten or more years of experience shipping production software. Someone who has owned a test harness at scale. Ideally, someone who has done that in a regulated environment. But at minimum, it should be someone who has worked in an environment where production defects had real consequences.

They need to understand software. They need to understand verification. They need to be able to read Gherkin and spot the scenario that is missing. They need to know when a contract looks complete but still leaves room for a bug to slip through.

And they need one more thing.

They need the backbone to sign their name to the release decision and mean it. At 6 p.m. on a Friday. With the CEO asking for the release. With the delivery lead pushing hard. With the team exhausted. With everyone wanting the answer to be yes.

That is the hardest qualification. It is also the one that matters most when the pressure is real.

Most organizations already have two or three people who can do this job. They are not always in QA. In fact, QA is often where these people end up when the organization has failed to promote them properly.

Find them. Promote them. Give them the reporting line.

Then trust them to do the job.

5. The first-pass verification rate

Of every number a pod tracks, first-pass verification rate is the one that catches the most expensive failure mode before it reaches production.

The expensive failure mode looks like this. The work feels productive. The harness output is mostly green. The team is shipping at a healthy clip. And underneath that surface, a regression is building that nobody will see until a customer trips it three weeks later, in production, in a way that is now twice as expensive to fix as it would have been to catch at generation time.

A pod that watches this rate weekly catches the regression early, when the fix is cheap. A pod that does not watch it ships at speed for two months and spends the third month rewriting code that should have been right the first time.

That is why first-pass verification rate is the metric I want every verification meeting to start with.

The definition is narrow on purpose. First-pass verification rate measures the percentage of VOSes whose generated output passes the acceptance contract on the first run of the verification harness. No regeneration. No hand-editing. No second try that quietly fixes the prompt, context, or contract.

If the harness fails once and the Intent Engineer adjusts the context pack before running it again, that VOS does not count as first-pass. If the AI Orchestrator re-prompts the model with better instructions and the second generation passes, that VOS does not count either. Those may still be useful learning moments, but they do not belong in the numerator.

The metric is trying to measure one thing: did we give the machine enough precision, enough context, and enough constraint that it got the work right on the first attempt?

That is why the number matters.

Healthy values depend on how mature the pod is. In the first four weeks, a pod will usually sit somewhere between 30 and 50 percent. That is not a crisis. That is the pod learning what its acceptance contracts actually need to say. The failed first passes are not wasted effort. They are where the Intent Engineer and Verification Owner discover the assumptions the contracts were carrying silently.

By week twelve, the rate usually climbs into the 50 to 70 percent range. The contract templates have started to stabilize. The Intent Engineer has the review rubric in their head. The pod knows more about what the machine needs before generation begins.

By year one, a healthy pod is usually in the 70 to 90 percent range.

Rates above 90 percent are possible, but I treat them carefully. When I see a pod first-passing almost everything, I do not celebrate immediately. I look for what is missing. A pod that passes every VOS on the first run is either unusually mature, unusually lucky, or quietly leaving scenarios out of the contracts.

The metric is useful in two ways.

First, it gives you a level reading. It tells you where the pod is on the maturity curve, which helps set realistic expectations for staffing, throughput, and planning.

Second, and more importantly, it gives you a trend reading. The direction of the rate week over week is an early-warning signal. A healthy pod's rate should climb slowly and then plateau. An unhealthy pod's rate usually moves suddenly, in either direction.

A sudden drop, meaning five or more points in a week, usually means one of two things.

The first possibility is that the context packs have gotten sloppy. New work introduced new assumptions, but those assumptions did not make it into the context the machine received. So the machine starts guessing. The acceptance contracts catch the guesses, and the first-pass rate drops.

That fix is usually mechanical. Audit the last ten context packs against the template. Find the missing piece. Restore it. Most of the time, that is an hour of work, and the rate recovers within a week.

The second possibility is more structural. The codebase may have moved underneath a recurring pattern. Maybe there was a refactor. Maybe a library changed. Maybe an architectural decision from another pod finally reached this pod's surface area. The machine is still generating against an older picture of the system, so the output starts failing in places that used to be safe.

That fix takes more work. The root context pack needs to be updated, and the acceptance-contract templates often need to change with it. If the Verification Owner catches the problem within a week, it may be half a day of work. If the issue runs for a month before anyone notices, it can easily become a week of cleanup.

A sudden rise is actually more worrying.

If the rate jumps more than ten points in a week, something probably changed. First-pass rates do not usually improve that quickly on their own. When they do, I assume someone has changed what "first-pass" means.

Maybe the Intent Engineer is pre-running the harness locally and revising the contract before submission. Maybe the Verification Owner is treating real harness failures as "harness issues" instead of VOS issues. Maybe the harness itself has been weakened.

Those are different versions of the same problem: the number has become the goal instead of the signal.

So when the rate jumps, investigate before celebrating.

I have the same conversation with almost every team in month two. Someone wants to tie first-pass verification rate to a bonus.

I always tell them no.

The rate is diagnostic. The moment you incentivize it, people will optimize for the number. Once that happens, the pod loses the signal and losing the signal costs far more than the bonus could ever save.

Read the rate. Act on the rate.

Do not pay for the rate.

6. What Verification Owners do not do

A Verification Owner is not a tester.

That distinction matters. Tests are produced from the acceptance contract. The Verification Owner sets the standard and reviews the contracts. The machine writes the tests from those contracts. The harness runs the tests. At no point in that process should the Verification Owner be manually executing test cases.

If they are, something is wrong. It usually means the harness has a gap. And the answer is not to turn the Verification Owner into a human safety net. The answer is to fix the harness.

A Verification Owner is also not a process enforcer.

The process should enforce itself. The VOS state machine should prevent a VOS from moving to SHIPPED unless the Verification Owner's signature is present. That signature is mechanical. It is attached to the evidence that the harness passed, the contract was satisfied, and the required checks were completed.

The Verification Owner's job is not to chase developers around and make sure they followed the steps. Their job is to protect the standard the signature represents.

If developers are trying to skip the steps, the state machine should catch it. And if it catches that behavior so often that the Verification Owner spends all day rejecting VOSes, the problem is not simply the developers. It is also not solved by lecturing people harder. The standard may be unclear. The workflow may be too awkward. The templates may be missing something. At that point, revisit the standard.

A Verification Owner is not a blocker of last resort.

If the only thing standing between a bad VOS and production is the Verification Owner saying no at the final gate, the pod has already failed upstream. At that point, the team is depending on one person's judgment instead of depending on the verification system.

Judgment matters. But judgment does not scale.

In a mature pod, the Verification Owner should be able to approve most VOSes quickly. The acceptance contracts are clear. The harness is green. The rollback path is ready. The compliance extensions have done their work. There is nothing dramatic to argue about.

If the Verification Owner is blocking two shipments a week, that is not a sign of a strong verification culture. It is a sign that verification is happening too late.

Move the problem upstream. Look at the contracts. Find the scenario that let the bad behavior through. Find the assertion that was too vague. Find the missing constraint. Tighten the contract, and the last-minute blocks should start to disappear.

A Verification Owner is not a compliance officer.

Compliance extensions are what enforce compliance inside the CID pipeline. The Verification Owner configures those extensions and audits whether they are working. But the person who interprets the regulation and writes the rule into the extension is usually someone from risk, legal, or the compliance function. Chapter 11 goes deeper on that collaboration.

The Verification Owner's job is to make sure the extensions are installed, active, and capable of producing blocking findings when a rule is violated. They are not the regulator. They are not the legal interpreter. They are the owner of the verification system that makes those rules executable.

These boundaries exist because I have seen each one violated in real engagements.

And the role does not survive those violations well.

A Verification Owner who becomes a manual tester is too busy to review contracts. A Verification Owner who becomes a process cop becomes too adversarial to be trusted with the signature. A Verification Owner who becomes the last-resort blocker burns out.

Protect the scope. That is how the role stays useful.

7. The handoff

The rest of the book rides on this chapter's discipline. The funding model in Chapter 9, the four metrics in Chapter 10, the compliance architecture in Chapter 11, the anti-patterns in Chapter 12, and the adoption path in Chapter 13 all assume the acceptance contract is the artifact, the harness is the judge, and the Verification Owner's signature is the trust anchor. Strip any of those out and the rest of the argument collapses. Keep them in and the argument is already holding most of its own weight. The reader who has gotten this far has the verification layer. The next chapters add the architecture around it.

Field Note from the Test Pilot

Casey Robinson, Intent Engineer

Where verification disagreed with me on SkipDay

When we started building SkipDay, our fitness app for everyday people, it was my first time creating a mobile app. I'd been writing VOSes on The Trade Codex for a while by that point, but SkipDay was a different kind of product, with a different audience, a different platform, and a different set of concerns. One of the early VOSes we reviewed as a pod taught me something about how the verification process works that I think is worth sharing.

The VOS was for the daily check-in flow. SkipDay's whole philosophy is permission-first. "Some is better than none, and none is OK too." The daily check-in is the flow where the user tells the app what equipment they have today, how much time they have, and whether anything hurts. If they mark a pain area, say lower back or knees, the app is supposed to build a workout that avoids exercises that would aggravate it and includes some targeted mobility work instead.

I had written four scenarios in the acceptance contract. By the next day, Glenn and David both had thoughts on three of them. Two of those I could see immediately once they pointed them out. The third became a conversation, and that conversation is worth telling because it shows what the verification process is supposed to do when people on the pod see the same feature from different angles.

The first scenario was about what happens when a user marks a pain area. I had written something like, "When the user identifies lower back as a pain area, then the workout avoids lower back exercises." David's feedback came back quickly: "What does 'avoids' mean here? Does the exercise get removed entirely, or does it get replaced with a modification? If someone marks their knees, do they lose all leg work for the day?"

He had a point I hadn't thought through. I had been thinking about it from the user's perspective. If your back hurts, don't make it worse. But I hadn't specified what the app should do instead. Remove the exercise and leave a gap? Swap it for something else? Include a modified version? Those are three different behaviors, and my contract didn't say which one.

So I improved it. I changed the contract to say the app should first look for a modification of the exercise that accommodates the pain area, and if no modification exists, substitute a different exercise targeting the same muscle group that doesn't involve the pain area. That was the better contract. David's feedback had found the gap between what I meant and what I'd actually written down.

The second scenario was about the mobility routine. My original contract said that when a user marks a pain area, the app should include targeted stretching and mobility work for that area. Glenn's feedback was short: "What if they mark three pain areas? How long does the mobility section get? Does it eat into their workout time?"

That was a good catch. SkipDay asks users how much time they have. 15 minutes, 30 minutes, 45 minutes, and so on. If someone marks three pain areas and the mobility routine for each one is five minutes, that's fifteen minutes of stretching before the main workout even starts. On a 30-minute session, that's half the time gone. The user came to work out, not to stretch for fifteen minutes.

I hadn't thought about the time constraint interacting with the pain area constraint. That's the kind of thing that seems obvious once someone says it, but I genuinely hadn't connected those two pieces of the feature in my head when I wrote the contract.

The third scenario was the one that turned into a real conversation.

My original contract said that the workout generator should always produce a workout, even when the constraints are tight.

My thinking was simple: SkipDay's whole brand is "we'll always give you something to do." An empty workout screen would feel like the app gave up on you. That's the opposite of what the product is about.

Glenn's perspective was different. He pointed out that if a user selects "bodyweight only" as their equipment, marks knees, lower back, and shoulders as pain areas, and says they have 15 minutes, the constraint space might be so tight that the generator literally cannot find enough exercises that satisfy all the conditions. His concern was: what does the contract say happens then? My contract just said "always produce a workout." It didn't say what to do when the math doesn't work.

Both concerns were real. I didn't want the app to show an empty screen. Glenn didn't want the contract to promise something the system couldn't always deliver.

We talked it through on a call, David and Glenn and me. The conversation was about ten minutes. We landed on something that addressed both sides: the generator produces a shorter workout when constraints are tight, with a warm message explaining the adjustment. Something like, "Today's a lighter day based on how you're feeling, and that's totally fine." The app never shows an empty screen. But the contract doesn't promise a full-length workout when the constraints make that impossible. It promises the best workout the constraints allow, with a message that keeps the tone right.

That scenario shipped. It's still the one that runs today. And it's come up more than I expected. Users do mark multiple pain areas, especially on days when they're sore from a previous workout. The warm message has been one of the most-commented-on pieces of the app in user feedback. People notice when an app acknowledges their situation instead of ignoring it.

That is the part I want to share about this experience.

I still think my instinct about never showing an empty screen was the right product instinct. Glenn still thinks the contract needed to handle the constraint-collision case explicitly. We both have a point. The better contract came from holding both concerns at the same time and finding the version that satisfied both.

That is what the verification process is for. Not to catch mistakes in a pass/fail sense. To give the pod a productive place to work through different perspectives on the same feature. The acceptance contract is where those perspectives get captured, and the contract that comes out the other end is better than what any one of us would have written alone.

I wrote a lot of contracts over the hundred days. About a third of them came back with feedback on the first look from Glenn and David. Most of the time, their feedback found real gaps, things I hadn't connected, edge cases I hadn't imagined, constraints I hadn't thought about interacting with each other. A handful of times, I had product context or user perspective they didn't have yet, and we worked through those together.

The contracts that came out the other end were better than anything I would have written alone. And that didn't require all three of us to agree on everything. Which was good. Because we didn't agree on everything. But we always found the version that was stronger than any one perspective on its own.

CHAPTER 7: THE CID PIPELINE IN MOTION

---

1. Forward flow, two eddies, parallel watching

The unit of work inside CID is the VOS. Not the sprint. Not the story. Not the release. A VOS is a single Verifiable Outcome Slice, and it makes one trip through the pipeline.

The pipeline has six states: DRAFTED, CONTRACTED, QUEUED, GENERATING, VERIFYING, SHIPPED. Forward flow, left to right. A VOS enters at DRAFTED. If everything goes well, it exits at SHIPPED, and SHIPPED is terminal. The VOS does not come back around.

There are two reverse edges, and only two. Both leave VERIFYING. The first goes back to GENERATING when the contract was right but the generation missed it. The second goes back to DRAFTED or CONTRACTED when the contract itself was wrong. Those are the eddies. They exist so the pipeline can correct fast when verification surfaces a gap.

They do not turn the pipeline into a loop. A VOS that takes an eddy still makes a one-way trip when it eventually reaches SHIPPED.

Observation is not a stage in the lifecycle. Observation is a watching layer that runs alongside the pipeline, in parallel, against the entire population of shipped VOSes already in production. When observation sees something that matters, it doesn't push the old VOS back to DRAFTED. It generates a new VOS that enters at DRAFTED and makes its own one-way trip. That is the kaizen channel. It connects production back to intent at the system level, not at the VOS level, and the connection is asynchronous and stochastic. Most observations don't trigger a new VOS. Most new VOSes don't trace cleanly to a specific observation. The signal is real, but it is upstream-flavored, not feedback-flavored.

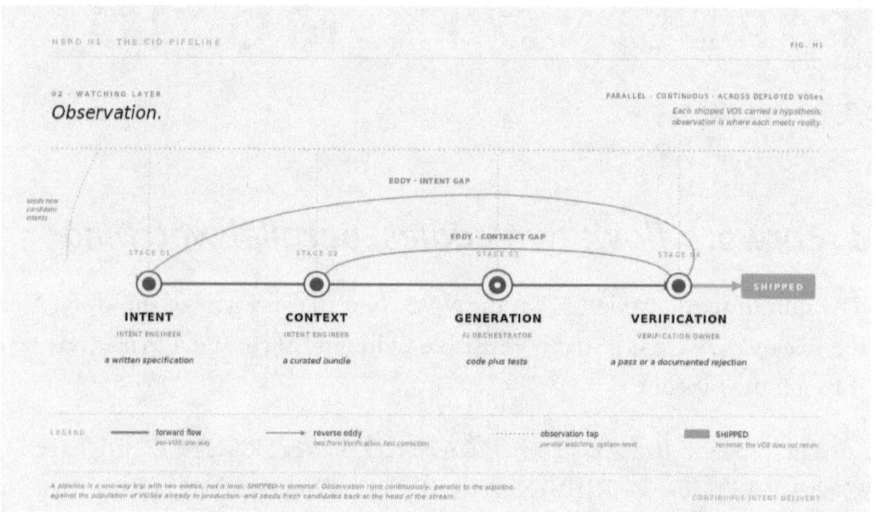

The CID Pipeline. Six states, forward flow with two eddies, parallel watching.

That is the geometry. The next several sections walk through what happens inside it.

The pipeline is also a learning system, but the learning happens at the system level, not inside any single VOS. The pod ships a VOS. The pod ships another. The fortieth VOS is faster than the fourth, because the context pack template has sharpened, the acceptance contract rubric has tightened, and the Intent Engineer has internalized two dozen Gherkin patterns that no one had to write down. The pod's velocity compounds because the pipeline's inputs are getting better, not because any individual VOS is being looped back through. A pod in its twelfth week is three or four times faster than the same pod in its second week, not because the people got faster at typing, but because the upstream stream of intents is getting cleaner.

A week inside a mature pod looks ordinary from a distance. The Intent Engineer authors one or two VOSes on Monday. The AI Orchestrator runs generation against the harness on Tuesday and Wednesday. The Verification Owner signs VOSes as the harness clears them. Rework happens on Thursday if it needs to, and shipping happens Thursday if it does not. Friday is the Pipeline Review, which is thirty minutes on most weeks and ends early on the good ones.

If the preceding paragraph sounds unimpressive, that is the argument. A pipeline running well is boring. The drama is in the weeks where something goes wrong, and the system is built so the wrongness shows up early, in a place where it is still cheap to fix.

2. The adaptive workflow inside Generation

Generation is one of the pipeline's stages, and it is the stage with the most internal structure, and it is where most of a pod's daily work actually happens. Inside Generation there are three phases and three depths. That is a three-by-three grid, nine cells in total. Every VOS picks a depth for each phase at authorship time, which traces a path through the grid and sets the protocol the generation run will follow.

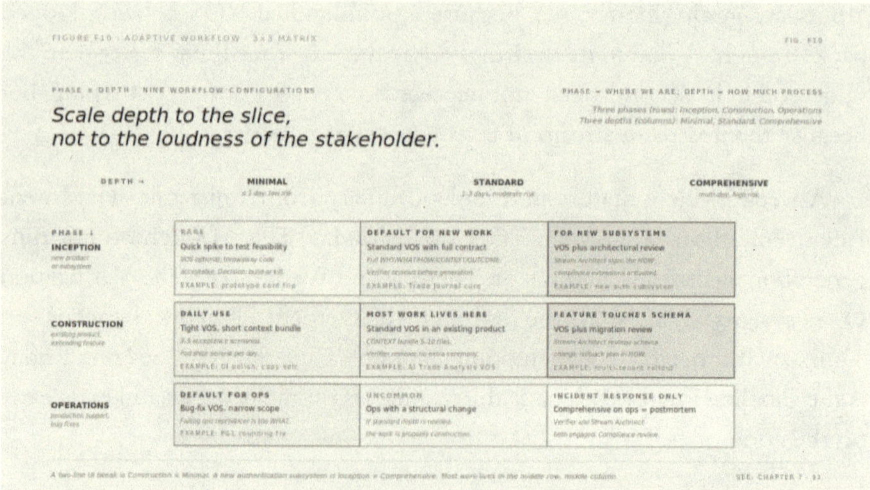

Figure F10 · Adaptive workflow 3×3. *Phase (Inception / Construction / Operations) by depth (Minimal / Standard / Comprehensive).*

The three phases: Inception, Construction, Operations.

Inception is the pre-generation phase. Requirements are validated against the acceptance contract, schemas are checked against the context pack, and the harness runs a dry pass to confirm the contract's scenarios parse.

Construction is the generation itself. This is the TDD red-green-refactor cycle described in Chapter 6, run one sub-task at a time, with user approval between parent tasks.

Operations is the post-generation phase. The release dry run, the rollback test, and the first seventy-two hours of monitoring after the VOS ships.

Three phases, in sequence, for every VOS.

The three depths: Minimal, Standard, Comprehensive.

Minimal is the depth a low-risk VOS runs at. A copy tweak on a marketing page. A label change. A color token swap. These do not need a full context pack. The acceptance contract is two or three scenarios. The harness runs visual-regression plus basic accessibility. Inception takes ten minutes. Construction takes twenty. Operations is a quick production spot-check. The whole VOS is done before lunch.

Standard is where most feature work sits. A new endpoint, a new form, a new report, a new screen. The context pack is five to fifteen files. The acceptance contract is eight to fifteen scenarios. The harness runs the contract plus static analysis plus security scanners. Inception is an afternoon. Construction is a day. Operations is seventy-two hours of monitoring plus a rollback drill. Most VOSes are Standard.

Comprehensive is the depth reserved for the things that break loudly. A payment flow. A PHI access path. A multi-tenant data migration. Something with a regulator attached. The context pack includes the full domain model and the compliance extensions that apply. The acceptance contract runs twenty or more scenarios, including the denied-path and failure-mode scenarios the rule exists to cover. The harness runs the contract, plus static analysis, plus security scanners, plus the compliance extensions, plus the threat-model checks. Inception is two days. Construction is two to four days. Operations is a staged rollout with a signed-off rollback plan and a real monitoring window. A pod will ship one or two Comprehensive VOSes a month, and those are the ones the CTO will be reading the postmortems on if something goes wrong.

The grid is not a single matrix pick a pod makes once. Each VOS chooses its cell per phase, independently. A feature can run Minimal Inception because the requirements are obvious and the Intent Engineer already drafted the scenarios, then Standard Construction because the codebase has unusual seams, then Comprehensive Operations because the feature touches the payment flow even though the code change is small. The depth is scaled to the risk at each phase, not to the VOS as a whole. This is the part that takes new pods a month or two to internalize. The old world had one depth. CID has nine.

A worked example anchors the grid better than another paragraph of theory.

Take a Trade Codex VOS Casey shipped in March. The feature was the Dual Theme System, VOS #7. The Trade Codex has two audiences: serious traders who want a professional dark UI, and gamification-oriented users who want the RPG tavern experience. The VOS added the ability for users to switch between a Trader theme (dark, modern) and an Adventurer theme (tavern, parchment, fantasy typography). Every component in the app had to support both themes from day one.

Inception: Standard. The requirements were clear but the scope was broad. Casey drafted six acceptance-contract scenarios: theme selection persists across sessions, Trader theme uses dark professional styling, Adventurer theme uses tavern/parchment styling with fantasy-appropriate language, theme never returns null (sync-first resolution so the user never sees a flash of unstyled content), and CSS variables are set before first paint. The Verification Owner reviewed the contract and flagged the null-state scenario as the one that needed the tightest specification. Inception ran for an afternoon.

Construction: Standard. The context pack included the theme provider component, the CSS variables file, and the user profile where the preference is stored. Five files total. The harness ran the acceptance contract plus the static analyzer. Construction ran for a full working day, including one subtask of regeneration after the first pass didn't handle the sync-first resolution correctly on initial page load.

Operations: Minimal. The feature is a UI preference. No regulated data. No financial transactions. No compliance exposure. The rollback path was simple: revert to the default theme. The monitoring window watched for one signal, the theme adoption split between Trader and Adventurer, to confirm users were discovering and using the toggle. Operations ran for seventy-two hours of passive monitoring.

That is one cell per phase. Standard Inception. Standard Construction. Minimal Operations. The VOS shipped in about three elapsed days, because the risk profile was UI-only and the Operations depth reflected that.

The grid is what made that allocation legible. A pod that runs Standard on everything would have spent three days on Inception that did not need three days, and would have spent eight hours on Operations when the rollout should have run a hundred. Either error is expensive in its own way. The grid prevents both.

The selector is the Verification Owner, in consultation with the rest of the pod. The AI Orchestrator flags the depth for Construction, because they know what the generation model handles well and what it does not. The Intent Engineer flags Inception, because they authored the contract and they know whether the scenarios are already tight. The Verification Owner owns Operations, because they own the rollback path. Three inputs, one selection, documented in the VOS header before any generation runs.

One caveat worth naming. In a pod's first month, everyone runs Standard on everything. Minimal feels reckless. Comprehensive feels paranoid. By month three, the selector gets used the way it was designed to be used. By month six, the pod is running Minimal on maybe a third of its VOSes and shipping two a day as a matter of course. The compounding from calibrated depth is one of the biggest sources of throughput gain after the first quarter.

3. Incremental generation and git discipline

The protocol that keeps the pipeline running safely at speed is deliberately unromantic. One sub-task at a time. User approval between parent tasks. A clean working tree as a prerequisite for every generation run. Commits authored by the machine, signed off by the AI Orchestrator, verified by the Verification Owner. The git history becomes the observation artifact.

Start from the clean-tree rule. Before any generation run, the pod runs git status, and the tree must be clean. No uncommitted changes. No stray files. No in-flight edits sitting in the working copy. If the tree is dirty, the pod stops. The prior task's changes get committed, reverted, or stashed with a named stash the Orchestrator writes down. Then, and only then, the next generation run starts. The reason for the rule is mechanical. The machine is about to make changes to the tree, and if the tree already has changes in it, the commit that captures the generation will capture the human's in-flight edits along with it. The audit trail is corrupted before the run finishes.

The next rule is one sub-task at a time. The VOS task plan decomposes a feature into parent tasks (1.0, 2.0, 3.0) and sub-tasks (1.1, 1.2, 1.3). Generation runs against one sub-task. The machine produces the test, runs it red, implements the minimum, runs it green, refactors. The sub-task gets marked [x] in the VOS document. Then the pod moves to the next sub-task, which is the next sub-task and not a batch of the following three. This is the rule new adopters push back on hardest, because the machine feels capable of doing more, and doing more in fewer steps feels like speed. It is not. Batching sub-tasks turns the generation into a black box the verifier cannot audit. When the batch fails, the Orchestrator cannot tell which sub-task introduced the failure, and the whole batch has to roll back.

The third rule is user approval between parent tasks. Sub-tasks (1.1 through 1.4) proceed without separate approval. Parent task boundaries (1.0 to 2.0) require an explicit green light from the human in the loop. The green light is not ceremony. It is a one-line check. The Orchestrator reads the diff for Task 1.0, confirms the tests pass, confirms the commit is clean, and types proceed. Then 2.0 starts. The approval gate is also the point at which context is reset, if the pod is working with a model that supports context reset. The gate keeps the generation window from drifting across unrelated parent tasks. That drift is the failure mode that produces code that looks right and fails for reasons nobody can trace.

Commits are the observation artifact. Every parent task that completes becomes a commit, messaged with a consistent template: VOS-<number> Task <task-number>: <brief description>. Every VOS that ships becomes a commit: VOS-<number>: Shipped: <VOS title>. The git log, read after the fact, is a complete operational record of the pod's work for the week. No meeting notes needed. No status report needed. The log is the record. When the Pipeline Review runs on Friday, the four numbers on the table are computed from the log plus the harness output. Nothing else is consulted.

The mocking rule is a sub-rule of TDD, covered in Chapter 6, but it applies inside every generation run and is worth restating. Mocks are permitted for external services only. Never for internal modules. Never for the database. The reasoning is the same as it was in the verification chapter. A mock of your own module commits you to a seam you may not actually have. If the machine invents the seam to make the mock work, the test is verifying the machine's invention instead of the system. Use the real database in the test harness. Use the real internal modules. If the test is too slow, the test harness has a problem the mock was hiding, and hiding it costs more than fixing it.

Working tree discipline. Incremental generation. Parent-task approval. Commit templates. Real-database testing. Five small rules. Every one of them has been violated in some pod I have worked with, and every violation has produced the same class of failure: generation that looks right, ships, and breaks a week later in a way the team cannot trace back to a specific change. The rules are not preferences. They are the minimum protocol that keeps the pipeline auditable at AI speed.

4. The pod, in three configurations

Casey, co-leading this section because the pod is the part of the methodology I have actually lived in, across three configurations, over fifteen weeks.

The pod is three roles. Intent Engineer, AI Orchestrator, Verification Owner. The three roles don't have to be three people. They have to be three functions, covered.

A pod has what *Team Topologies* calls a team API. The surface the rest of the organization interacts with is small and cleanly shaped. Intent comes in one door. A signed, shipped VOS goes out the other. Everything else about how the three roles coordinate internally is the pod's business, and the rest of the org does not need to watch it happen.

Alchemaize ran three configurations over the 100-day sprint. The 1-FTE pod (me, on TradeCodex and then early Finaize, the first few weeks). The 2-FTE pod (David and me, with Glenn fractional on verification, weeks four through nine). The 3-FTE pod (all three of us, one role each, weeks ten through fifteen). I'll walk through each one because the ergonomics are different, the failure modes are different, and the reader trying to figure out which configuration to run should know what each one actually feels like from inside.

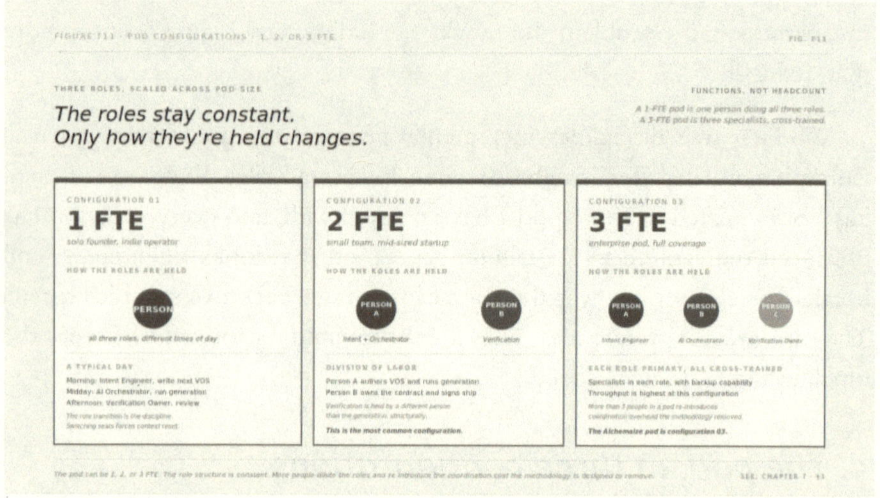

Figure F11 · Pod configurations 1·2·3 FTE. *Role coverage across 1-FTE, 2-FTE, and 3-FTE pods.*

1-FTE pod

One person, all three roles. This is the smallest viable configuration, and it is more intense than it sounds on paper.

I ran this for a few weeks across TradeCodex and early Finaize. The calendar looked like this: Monday morning I wrote the VOS. Monday afternoon I curated the context pack and drafted the acceptance contract.

Tuesday morning I ran generation against the harness. Tuesday afternoon I verified my own output, which is the part that is hard. Wednesday morning I shipped what worked and queued the rework on what didn't. Wednesday afternoon I started the next VOS.

Two things about that schedule. One, it's doable. I shipped several VOSes across TradeCodex and early Finaize in those weeks, and both apps are in production. Two, it's fragile. When one role fails, all three roles fail, because the one role is me. I got tired. I missed a scenario in the acceptance contract on one VOS that my own verification pass didn't catch because my own intent pass had missed it first. The bug shipped. It wasn't bad, but it wasn't good either, and a bigger product would have made me pay for it.

A 1-FTE pod works for small streams. A single-purpose tool. A marketing site. An internal utility. It does not work for anything where the cost of a missed scenario is high. And it does not work long. A few weeks was the ceiling for me. By week four I asked David if we could bring him in on AI Orchestrator, and we did, and the 2-FTE pod started.

Common failure modes, in order of how often I saw them. One, verification fatigue. The same person who wrote the contract cannot catch the thing the contract was silent about, because the silence is in their head too. Two, context collapse. When you are also doing generation, you stop curating context and start throwing the whole repo at the model, which produces worse output and trains you to trust the model less. Three, schedule compression under pressure. When the week gets tight, the discipline gets cut, and the discipline is what was making the speed safe.

If you are going to run a 1-FTE pod, do it for a bounded stretch and do it on a small surface. Then scale up.

2-FTE pod

Two people, three roles. One of the people wears one role full-time; the other person wears the other two. The standard Alchemaize shape for weeks four through nine was me on Intent Engineer full-time, David on AI Orchestrator full-time, and Glenn on Verification Owner at maybe 30 percent, covering the verification signature and the harness work without being in the pod's daily standup.

This is the configuration most readers will end up running. It is the configuration that matches the economics of a small team that has one senior verifier shared across multiple pods. It is also the configuration the Alchemaize founding team ran longest, and it is where we developed most of the practices that are now in the methodology.

A week in a 2-FTE pod looks like this.

Monday morning, I write the Monday VOS. Intent and context, drafted in my notebook and then transcribed into the VOS template. I slack David around eleven with the draft. He reads it over lunch, flags the files I missed in the context pack, flags the one scenario that's ambiguous in the acceptance contract.

Monday afternoon, I fix the draft. David starts the Inception dry run against the harness. Glenn reads the contract async, sends two notes, one of which is right and one of which I push back on, and we resolve both before 4 p.m.

Tuesday, Construction. David drives generation. I sit next to him on video for the first two sub-tasks, because the first two are where the model's interpretation of the contract is visible, and if the model is off I want to see it early. After sub-task 2.0 we break, and I go write the Tuesday VOS while David runs sub-tasks 3.0 through 5.0 on the first one. Glenn signs the parent-task commits as they clear the harness.

Wednesday morning, Tuesday's VOS ships. David moves to Construction on Tuesday's VOS. I finish writing the Wednesday VOS and start the Inception dry run. Glenn spends thirty minutes in the harness looking at the compliance extension output on Monday's VOS, because Monday's VOS touched a payment flow and he wanted a closer look at the denied-path scenarios.

Thursday is either ship or rework. Wednesday's VOS clears the harness by Thursday morning if Inception was clean. If not, Thursday is the day I rewrite the ambiguous scenario, David re-runs generation against the revised contract, and we ship Thursday afternoon. The rework day is not wasted. It is the day the pipeline is catching a problem before the problem reaches production.

Friday is the Pipeline Review in the morning and the next Monday's VOS Intent work in the afternoon. Friday is also the day Glenn pulls the week's harness reports and writes the one-page summary that will be on the table at the review.

Three VOSes a week is a realistic cadence for a 2-FTE pod running at maturity. We ran at four on some weeks and at two on the weeks where one of the VOSes was Comprehensive. The pace varies. The cadence doesn't.

Common failure modes. One, the fractional Verification Owner gets pulled into other pods and their attention on your pod drops. When that happens, the verification signature starts to look ceremonial rather than real, and the pod has to decide whether to block shipping or accept signatures that haven't been earned. Block. Two, the Intent Engineer and the Orchestrator become one person's workflow over time. I did this with David twice. I wrote the VOS the way I knew he'd run it, and David ran it the way he knew I wrote. The pipeline looked efficient and was actually a closed system that the verifier couldn't catch, because the Intent and Generation stages had merged in our heads. The fix is to rotate. David wrote VOSes for two weeks in February. I ran Orchestrator for a week. The rotation breaks the merge.

3-FTE pod

Three people, one role each. This is the mature configuration the enterprise layer assumes. It is also the configuration that produces the highest throughput and the cleanest verification signal.

I have lived it for five weeks now. The week is quieter than the 2-FTE week. Not because there is less work, but because the work is parceled cleanly and nobody is wearing two hats. I wake up, write intent, hand off, write the next intent, hand off. David orchestrates. Glenn verifies. Each person is doing one thing at a time, which is the condition under which the thing actually gets done well.

Monday I write two VOSes. Tuesday I write another one and review Glenn's feedback on Monday's. Wednesday I write the Thursday VOS and sit with David for fifteen minutes on the Wednesday generation run because he has a question about a scenario. Thursday I write the Friday VOS. Friday morning is Pipeline Review. Friday afternoon I read the next week's product roadmap and draft intent for the first two VOSes of Monday.

David's week is symmetric. He orchestrates four generation runs on average. He owns context curation for each one, which he has developed into an art form I am not going to pretend I can describe accurately. He runs the Inception dry runs. He sits the parent-task approval gates. He commits the work.

Glenn's week is different because Glenn's time is not bound to the daily cadence the way Intent and Orchestration are. Glenn signs VOSes as they clear the harness; on a good week the signatures are a few minutes of work per VOS. Glenn also owns the harness itself, which means when a new compliance extension has to be added or a new static analyzer has to be integrated, that work is Glenn's, and it fits around the signature work. Glenn is also the person who reads the week's git log end-to-end on Friday morning before the Pipeline Review, which is the single most useful practice for catching patterns nobody else would see.

Throughput on a 3-FTE pod, at maturity, is five to eight VOSes per week depending on the depth mix. Our record week was twelve, and that was a week with nine Minimal VOSes on a content-site redesign.

Common failure modes. One, role creep. The Intent Engineer starts reviewing the harness output, the Orchestrator starts drafting acceptance contracts, and the Verification Owner starts suggesting scenarios before the contract is written. Each one feels collaborative and each one degrades the separation of concerns that the three-role structure exists to protect. The rule I have learned to hold: collaborate in Inception, then stay in your role through Construction and Operations. Two, the pod outrunning the product. When the pod is at full 3-FTE maturity, it can ship faster than the product owner can decide what to ship. The bottleneck moves upstream. This is a good problem to have, and it is also the moment the enterprise layer has to show up, which Chapter 14 covers.

Three configurations, three calendar patterns, one pipeline. The differences are real and the pattern is the same: intent early in the week, construction through the middle, ship or rework Thursday, review Friday. Scale the depth, not the discipline.

5. The Pipeline Review

The Pipeline Review is thirty minutes, weekly, Friday morning in Alchemaize's case, on whatever day the pod settles into. It is the only standing meeting on the pod's calendar.

Four numbers on the table. Cycle time, first-pass verification rate, outcome KPI delta, cost per shipped intent.

Cycle time is the median wall-clock time from Intent authored to VOS shipped, measured across the week's shipped VOSes. A mature 2-FTE pod runs at 2 to 3 days of cycle time on Standard VOSes. A 3-FTE pod runs at 1 to 2 days. Minimal VOSes pull the median down; Comprehensive VOSes pull it up. The number is a level reading and a trend reading. A sudden jump in either direction gets a question in the review.

First-pass verification rate is the metric from Chapter 6. Percentage of VOSes that pass the harness on first run, no regeneration, no hand-editing. Mature pods run between 60 and 85 percent. The trend matters more than the level. A sudden drop is the pod's early-warning signal.

Cost per shipped intent is the pod's fully loaded weekly burn divided by the number of VOSes shipped that week. The number is blunt. It is also the most honest metric the pod has. A week where the cost doubled is a week where the pipeline was not running efficiently, whatever the narrative around it. The Orchestrator who built this number first can explain exactly what drove the spike; the Orchestrator who can't has not been reading the telemetry.

That is the agenda. Four numbers, the trends across the last four weeks, and two questions: *is anything flagged, and does anyone see a pattern that needs a decision?* If the answer to both is no, the meeting ends early. On good weeks it ends in ten minutes. On harder weeks it runs the full thirty and occasionally a bit longer, because a real pattern has surfaced and the pod needs to decide what to do about it.

No demo. No status report. No slide deck. The git log and the harness output are the record. The review is a signal check, not a performance. When a pod starts dressing up the Pipeline Review with demos or narrative summaries, the meeting has stopped serving its function and started serving an audience, and the first thing to audit is whether the audience has any business being at the review in the first place.

The brevity is the argument. A pipeline that is running well produces a review that is short. A review that is long is telling you the pipeline is not running well. Listen to what the meeting's length is saying about the work underneath it.

Field Note from the Test Pilot

> *Casey Robinson, Intent Engineer*
>
> **What a Wednesday looks like**
>
> Wednesday, March 4. Claremore. Early spring rain on the windows, that Oklahoma kind where it's starting to warm up but the sky hasn't gotten the message yet. I can hear it from my desk, which I like.

7:15 a.m. Up. Coffee. I head to the kitchen and start breakfast before my son comes in. Eggs and toast. He shows up a few minutes later with *The Champion's Mind* tucked under his arm, the one about how great athletes think, train, and thrive. He's been reading it for a couple of weeks now. Spring football is right around the corner and he's getting his mind right for it. He's pretending to read but is actually eyeing my plate. I give him the toast. He takes it without thanks, which is how it should be. We sit together for a few minutes before he heads off to school.

7:45. Back at my desk. I open the laptop, check Slack, pull in the latest changes on The Trade Codex and SkipDay. Nothing urgent overnight.

8:00. Daily call with David and Glenn. I want to describe this call honestly, because it's the only meeting we have every day and it's not what most people would expect from a methodology book. It's a call amongst friends. We talk about life, family, what's going on in the news. Glenn shares something his family did last night that has all three of us laughing. David and I talk about a couple of pre-market setups we're watching, all three of us are traders, so we look at opportunities together, share strategies, debate whether a move is real or noise. Today there's an interesting setup on a semiconductor name that David spotted and we spend some time on it. Work comes up if it comes up. There's no agenda. Sometimes we talk about a feature or a VOS for a while. Sometimes work doesn't come up at all and we just catch up on life. Today, somewhere in the middle of the trading conversation, I mention I'm going to write the SkipDay onboarding VOS and also knock out a small Trade Codex VOS. David has a thought about the SkipDay onboarding, what happens when a user skips the workout style selection. We talk it through for a few minutes and land on a good approach.

8:45. The call is still going, we're watching the market open together, but I start working on the Trade Codex VOS while we sit there. Users have been asking for the weekly performance report email to show their win rate alongside the P&L number, because a positive P&L with a low win rate tells a different story than a positive P&L with a high win rate. Small feature, matters to the people who use it.

9:00. Call wraps up. I finish writing the Trade Codex VOS. Intent is three lines. Acceptance contract has four scenarios: P&L with trades in the period, P&L with no trades, win rate calculation including only closed trades, and the formatting when the win rate is exactly 50 percent (which needs to show as "50%" not "50.0%" because the extra decimal looks weird in an email). I describe the context areas to my AI tool and it identifies the files.

9:15. I run it through generation. This takes a little while. I review the output, check the scenarios, make sure the formatting looks right. I send it to David on Slack for a quick look. He's good with it, suggests adding the win rate to the email subject line too. Good catch. I update the VOS, regenerate, review again. Ships clean around 10:15.

10:15. I let the pets out, make my wife a cup of coffee. I've been working on my latte art, trying to put a heart design on top. It's getting better. Not great yet, but better. She smiles. We talk for a few minutes.

10:30. Back at the desk. I read through the SkipDay exercise library spec to prepare for the onboarding and workout generator VOSes. I'm not reading code. I'm reading the product spec to understand how exercises are categorized and how the onboarding connects to the rest of the app.

11:00. Switch to SkipDay. I write the onboarding VOS based on what we discussed on the morning call. The WHY is about setting the permission-first tone from the very first screen. The WHAT has six acceptance criteria. I describe the context areas to my AI tool. The VOS takes about forty-five minutes to draft, then I run it through generation and review the output. The welcome screen tone is warm, the workout style selection works, the skip behavior defaults correctly. I send it to Glenn since he's been on the SkipDay architecture this week. He has one note: what happens if the user force-closes the app mid-onboarding? I hadn't thought about that. I update the contract, regenerate, review. Glenn's good with it. The whole thing from start to ship takes about two hours. Ships around 1:00.

1:00. Lunch. Grilled cheese on homemade bread. I step away from the desk and eat with my wife. We talk for a bit.

1:30. Back at the desk. I start drafting the SkipDay workout generator VOS, the feature that assembles exercises based on the user's equipment, time, and pain areas. This is a bigger VOS and I know the acceptance contract needs to be thorough. I work on it through the afternoon.

3:00. I finish the workout generator VOS draft. It's not ready to run yet. I want to sleep on the acceptance contract and review it fresh tomorrow morning. I save it.

3:15. Slack for a bit, checking in with David and Glenn on a couple of things. One of them says *thanks Casey*, and I note that the thanks is the first human thanks I've received all day, which is also the right amount.

3:30. I head out to pick up my son from school. That's the highlight of the day, every day.

4:00. Home. We hit the gym together, my wife and I, with my son getting ready for spring football around the corner. He's been putting in the work and it shows. We're there for about an hour.

5:15. Home, shower, start thinking about dinner.

6:30. My daughter calls from college. She tells us about her training for the day. We catch up on how things are going. These calls are one of the best parts of the week.

7:00. Dinner.

7:30. Shows until about 10. My wife and I on the couch. Nothing productive. Nothing that needs to be.

10:00. Bed. On to the next day.

Here is the tally for the work part of the day. Two VOSes shipped, one on The Trade Codex (the weekly report win rate), one on SkipDay (the onboarding flow). One VOS drafted for tomorrow (the SkipDay workout generator). Thirty minutes reading a product spec. An 8 AM call with David and Glenn that covered family, trading, life, and a little bit of work. Made my wife a coffee with a heart on it. Grilled cheese on homemade bread. Hit the gym. Picked up my son. Talked to my daughter. That is the day.

I want to be honest, this was probably on the lighter side. Some days there's more going on. A consulting opportunity comes in that gets all three of us excited and we spend the afternoon talking through the approach. Some days we really focus and push out more VOSes because the momentum is there and the features are flowing. The methodology supports both kinds of days because the VOS structure doesn't care whether you ship two or five. It cares that each one is clear, verified, and real.

Two things I want the reader to notice about that Wednesday.

First, I worked on two different applications in the same day. The Trade Codex in the morning, SkipDay through the rest of the day. The context switch between them was not painful. I wrote a VOS for one, described the context areas, ran it, reviewed it, shipped it. Then I wrote a VOS for the other. Each VOS took about an hour and a half to two hours from start to ship. The VOS structure makes the context switch clean because the VOS carries the context with it. I don't have to hold the whole application in my head. I hold the feature in my head, and the VOS holds the rest.

Second, nothing about the day was heroic. The VOSes that shipped did so because David and Glenn each reviewed one, because the acceptance contracts were specific enough that the AI produced clean output, and because the pod works as a team where we support each other. One of the things I look forward to most is the daily conversation, not because of the work, but because all three of us are in agreement on how to build together.

> There's no conflict. Just taking action to get things done. That's what makes the day frictionless. The methodology supports it, but the trust between the three of us is what makes it feel easy.
>
> I wrote this up because the methodology is often described by the extraordinary outputs it produces, and the temptation is to imagine the days that produced the outputs were also extraordinary. They aren't, and that's actually the interesting part.

6. What the chapter has told you

The pipeline runs forward. The watching layer runs alongside. The pod does the work. The adaptive workflow calibrates depth to risk. The git discipline keeps the record clean. The Pipeline Review checks the signal, weekly, in thirty minutes. Scale the depth, not the discipline.

The next chapter is Casey's solo field report from the test-pilot seat. The chapter closes the book's Practice section and hands off to the Enterprise section. Read it. He is the second protagonist, and the pipeline you just read about is the one he lived inside for fifteen weeks.

Notes

1. Matthew Skelton and Manuel Pais. *Team Topologies: Organizing Business and Technology Teams for Fast Flow.* IT Revolution, 2019.
2. Ken Schwaber and Jeff Sutherland. *The Scrum Guide,* 2020 edition. scrum.org/resources/scrum-guide (accessed 2026-04-22).

CHAPTER 8: FIELD REPORT FROM THE TEST PILOT

---

1. Who I am and what I'm not

Being from Oklahoma and a big sports enthusiast, I often get looked at as a retired football player more so than a consultant or even a technologist. Standing at six foot two inches and big frame I am often caught out on a football field somewhere coaching in my free time. That's relevant context, not a joke. When I walk into a meeting with David, the people in the room sometimes assume he's the consultant and I'm his security detail. The assumption's reasonable if all you have to go on is the way I look. I mention it because what I'm going to tell you in this chapter is that the thing I do for a living now is author the specifications for production software that runs against real customer data. If you'd told me on the day I was brought into Alchemaize that this would be part of my job, I'd have laughed and then looked at how I can share knowledge or help streamline processes for operations.

The actual job I was brought in for was to become the Chief Operations Officer. Supporting a startup that was aspirational in looking to build a product called Ember. My background is not in engineering or anything related to building a codebase or deploying infrastructure. Even though I was working for one the biggest cloud providers, my roles were more akin to consulting, customer service, operations, and Agile delivery, carried for long enough that it's the shape my career has. For four years immediately before Alchemaize I was at Amazon Web Services, at the Senior Manager level, leading a team of Customer Solutions Managers. If you haven't encountered that role before, a Customer Solutions Manager at AWS is the person who stands in the seam between an enterprise customer and the technical capability they're trying to adopt. You guide the migration. You hold the plan together. You translate between the customer's internal politics and the technology's actual shape. You do a great deal of it in Agile ceremonies, because that's the format enterprise delivery has mostly settled into. You don't write production code. That was my world for the immediate stretch before this one, and variations of it have been my world for most of my working life.

I say all that so the rest of the chapter lands cleanly. I've been in software-delivery rooms for over twenty years. I've run Agile programs. I've written requirements, backlogs, acceptance criteria, operating playbooks, and the kind of customer-facing documentation that has to survive a lawyer reading it. What I've never done, in any of those rooms, is write code. Not a little on the side, not as part of any technology implementations, none developed in my career. I can discern how technology supports a customer, I can explain it, but the under the hood interworking's to make it work is like a foreign language to me. I can understand what is broken or not working and what are some culprits are from the technology stack. But what I can't do is type the fix. The twenty years of enterprise-delivery rooms sharpened the first skill without ever teaching me the second one, and until January I'd thought of that as a ceiling.

I didn't come to Alchemaize planning to be anything other than the person who kept the company's operations intact while everyone else focused on building Ember. That was the plan when I was hired. It is crazy how plans change and evolve. What happened instead was a change in what a certain kind of operational person can credibly do at work. The change is less dramatic than it sounds, and it's at the same time the most consequential professional thing that has happened to me, and to put into words, is a liberation I never thought possible before.

And here's the fact this chapter is in the book to explain. In the hundred days covered by this book, I personally shipped production code, first on an app called The Trade Codex, then moved on to build Finaize, a modernized Auto F&I web app with a mobile companion, without David or Glenn touching the code I shipped. I did it under a methodology we created and evolved and had been building but hadn't yet named; the naming came after they watched me do it repeatedly. The methodology is what the rest of the book is about. This chapter is what it looked like from where I was sitting.

2. The week the floor came up to meet me

It was a Thursday at the end of January, January 29 to be exact, that David walked me through opening my IDE for the first time. To explain why that day became the day I finally tried writing code, I have to tell you about the day before it. The day before is the day everything I had assumed about the next stretch of my life turned out not to be true.

A few things had happened at the company first. Ember, the product I'd been hired to help launch, an AI-augmented reading application David and Glenn had founded Alchemaize to build, had shipped its MVP back on December 1. The product worked beautifully on public-domain books from Project Gutenberg, and what we'd discovered through December was that the audience the user-acquisition story depended on did not want public-domain books; they wanted Harry Potter and Fourth Wing and the books that are actually being read right now, and DRM made that catalog inaccessible to us without a Kindle integration we hadn't been able to start a conversation about. By the end of December the development team that had been helping us build Ember was wound down. The Series A push I'd been brought on to run was, by the second week of January, not the path forward.

David and Glenn had been experimenting through the fall with AI-augmented development (Cursor, Kiro, patterns of generation and review), and they always joked with me about hey it would be cool to teach you how to write code, it would be great to build something together and I always laughed it off and said "Yeah, maybe one day". By early January they'd reached a conclusion neither of them had expected to reach when the year began. The shape of software development itself was changing, at their desks, in real time, and Alchemaize was going to bet on the change rather than work around it. On January 18 we drew a line in the sand, named the new approach Continuous Intent Delivery, and agreed I was going to try writing a VOS the next time we had a video call free for it. The video call free for it landed ten days later, but a different thing happened in between.

On Wednesday, January 28, 2026, at four in the morning Central time, Amazon issued a company-wide reduction in force. Sixteen thousand Amazonians received notices that their positions had been eliminated, with a 90-day notification period starting that day. I did not see the email, because I was asleep. David did. David tried to access an Amazon resource, could not, realized what had happened, and texted me at 6:05 AM. *Hey, I think our whole team got hit including me.* The text sat on my phone for two hours.

I woke up at eight. The phone had a wall of messages on it. The first ones I read were from my own team members, the early-migrations and EBA folks I'd been leading inside David's broader org. They were all in the same shape. *Hey Casey, I was impacted by the RIF, hope you are okay, it was awesome working with you.* Every one of them was assuming I was safe, because as their manager I usually was. They were reaching out to say goodbye in case the morning ate the rest of the day. I read three of those texts before I got to David's, and the moment I got to David's was the moment I understood that I was not the manager checking on the team this morning. I was on the same list they were on.

I texted David back at 8:18. *Yep...* I have read those four characters back many times. They were the only thing I could type.

By 8:20 we were on a Zoom call. David and I have run a lot of Zoom calls together over the last twenty years, and this one was unlike all of them. It was not a working call. It was not a planning call. It was the two of us sitting on a video call trying to figure out how to be useful to fourteen people whose mornings looked the way ours did and who had been our teammates the day before. We spent most of that morning replying to texts, telling people what we knew and what we didn't, asking what they needed. By the end of the day David had set up a Discord channel for the full team, a place to communicate, vent, ask each other for leads, support each other while everyone figured out what came next. The channel is still active months later. I check in on it most weeks.

I want to say something about that morning that is not the part of the story that makes the methodology look heroic, because the methodology was not what I was thinking about that morning. I was thinking about my team, and about my mortgage, and about what I was going to tell my wife when she came down from upstairs and asked how the morning was going. The methodology was sitting in a folder on my laptop where David had told me we were going to write our first VOS later that week, and I will be honest, I had not yet decided whether I was actually going to try it. The night before I had told my wife I was going to try something new. That was Tuesday night, before the RIF. By Wednesday afternoon, the something-new had stopped being a thing I might do and started being the only thing I had to do, because the AWS job I would have been doing instead was now ninety days from being over.

There is a version of this story that says the layoff arrived as a gift, a forcing function that handed me the focus the methodology needed, and the universe rearranged itself to deliver Casey to his calling. That version is not honest. The morning was a blow. People I cared about lost work they were good at, and David and I were among them. What is also true is that the morning rearranged the question I had been carrying since January 18 from *will I try this* to *what else am I going to do.* Both of those are real. The book asked me to write the truth and the truth has both halves.

On Thursday morning, January 29, David and I were on another video call. He shared his screen.

The something different was that David and Glenn had been watching their own development speed up by an order of magnitude, and the speedup had surfaced a claim they couldn't yet verify. The claim was that the skills the speedup rewarded were not the skills the old methodology had rewarded. The bottleneck had moved from planning, typing and iterating on code, which is that foreign language I do not understand, to specification and verification, and a person whose career was about clarifying intent (like mine) rather than implementing it could, given the right framework around the generation step, ship production software. When I decided it was time to try, it was out of a genuine approach to see what I could do. To lean into the can I "really" do this even if I don't know what is going. My thought was I would need to lean on Glenn and David to show me and teach me what I need to do. They got me setup and shared some guardrails I need to be aware of and then let me go, checking in periodically which then turned into the focal point of our conversations every time we met.

I was curious, and I also wanted to be helpful, and I'd spent enough years at AWS watching enterprise customers struggle with the same specification-and-handoff problem that I already had a theory the handoff was the expensive step. Why, well the process of planning work, holding daily standups and timeboxing development created a wait time that in essence queued a season professional in development to pick up the work, that wait could be hours or it could have been days or potentially a week before it was ready to be picked up. This new approach, was to try this new way, with an IDE to help but reduce that wait time and take an idea to production to see how far I could get on my own. I assumed I'd probably fail, and we'd all learn something, and we'd go back to the Series A push. I was wrong about the first part of that assumption. The second part turned out to be true, though not in the way any of us expected.

I want to be honest about what I was thinking the night before that call. I had spent two decades being the operations person in software-delivery rooms, and the line between me and the engineers had always been a real thing. The engineers wrote code. I wrote requirements. The requirements got translated by the engineers, sometimes faithfully, sometimes not, and the back-and-forth on that translation was a big part of what I had been good at for twenty years. Walking across that line, even with David and Glenn telling me it was different now, was not nothing. I had the laptop open. I had the AI tool ready to go. I had told my wife I was going to try something new this week and she had said good luck and meant it. And I sat there for about ten minutes before the call started, just looking at the screen, asking myself if this was going to be the day I finally tried, or if this was going to be one more day I told myself I'd try later.

The first thing David did was help me set up my tool. This tool was the entry way for me to understand how to set up my project, put in the right guardrails to be able to ask the right questions and get connected to the pieces I needed to be successful.

What that setup actually looked like, on the screen, was simpler than I had imagined and at the same time required more discipline than I had expected. David shared his screen and walked me through opening my IDE for the first time. He showed me a folder structure I would later learn to take for granted: a `vos/` directory where the specifications would live, a `context/` directory for the curated files, a few configuration files that told the AI tool how to behave inside this project specifically. He walked me through a rules file that set guardrails. No production credentials in code. All database changes go through migrations. All generated tests run against the real database, not a mock. These were the kind of constraints I'd heard Glenn argue for in a hundred meetings without quite realizing those arguments were about to become the floor of how I personally worked.

Then he showed me a VOS template. Five sections, the same five David and Glenn had been working with for weeks. WHY, WHAT, HOW, CONTEXT, OUTCOME. He read each one out loud and said what each was for. The WHY was the business outcome, the user need, the reason this thing existed. The WHAT was the contract, written in a language called Gherkin that I would learn to write the way I had once learned to write Spanish in high school, slowly and with a lot of mistakes and eventually with enough fluency to argue with somebody in it. The HOW was the technical approach, which David said most of the time the AI Orchestrator, which would be him for my first few VOSes, would write. The CONTEXT was the four to fifteen files the AI needed to see. The OUTCOME was the measurable thing the VOS predicted would change, in dollars or in user behavior or in defect counts.

Then he closed the template and asked me what feature I wanted to build first.

Then the fun part, we worked through the first specification of what we wanted to do which spawned the VOS mentality. It wasn't me saying what the system is architected to do, it was me saying out loud what the outcome should be, what would be the acceptance of it, and how do I want it to behave. No code, no architecture, nothing of that nature, just what I wanted to drive from an outcome, in human to human speaking language, nothing more. We were very precise about the framing. We didn't just "write the requirements." We didn't go through the "describe the feature" exercise. It was what did we want to *write down that the application will do (outcome focused), in a way that is specific enough that a machine could not possibly misinterpret you, and test it by imagining the machine that could not.*

I'd been asked to write requirements documents before, in other jobs. I'd even been pretty good at it in the context I'd been asked to do it in, which was the context of writing requirements for human software engineers who would read the document, fill in the parts I'd been vague about from their own professional judgment, and produce something approximately correct. That's what requirements documents had always been for, in my professional experience. They were a rough sketch the engineer then did the real work to complete.

Even though it seemed similar, we were working on something different. We were designing the outcome of the ask of the system to be complete enough to stand alone. What we were doing was taking out of the document all the assumptions I'd normally have left for the engineer to fill in, and to replace every one of them with a sentence as guardrail or ask.

Here's the practical thing. The skill of writing a VOS is not the skill of writing code. It's the skill of knowing, precisely, what you want, and being willing to say it in a way that cannot be negotiated. Those are two distinct skills, and the second one is harder than it sounds, because most of us spend our working lives writing descriptions of things that are deliberately fuzzy so they can be negotiated later. Sometimes that's diplomacy. Sometimes it's leaving room for the engineer's judgment, or for a stakeholder's face. In most rooms, those are the right moves.

What a VOS asks for is the opposite of diplomacy. It asks for a specification that would leave no room for doubt and is more authoritative. *When the user identifies a pain area during the daily check-in, the workout generator shall avoid exercises that aggravate the identified area, substitute a modification where one exists, and include targeted mobility work for that area; any generated workout that includes an exercise contraindicated for a marked pain area is a verification failure and the workout shall not be presented to the user.* That's a sentence I wouldn't send in an email. I'd soften it. I'd phrase it as *we really need to make sure the workout doesn't include anything that would make their pain worse.* That phrasing would, in the old world, have been fine. An engineer reading it would have understood what I meant.

A machine reading it does not understand what I mean. A machine reads what I wrote. If I write *we really need to make sure the workout doesn't include anything that would make their pain worse*, the machine will produce something that is, on average, pain-aware-ish, and sometimes it will be and sometimes it won't, and when it isn't, the bug is my fault, not the machine's, because I didn't say what I wanted.

Writing a VOS is the practice of saying what you want and being specific about it.

The application I cut my teeth on was a small one: The Trade Codex, an application to help somebody interesting in investment trading, learn to do just that, place investment trading. I have spoken with so many people that they really don't know how to trade and it is an unknown for them. So what did I do about it? I created an application that was a journal first, uses AI to provide feedback on your trades you make, along with bringing in a learning path to learn the basics of trading. All of this while bringing in my own flavor of a fantasy world to help bridge the gap between trading and games that I enjoy.

I wrote four VOSes against TradeCodex over a weekend. It was the training ground. Something I thought impossible started to turn into something realistic. I was able to put AI to work, define my intent and what I wanted to accomplish, and AI orchestrated the whole build with me overseeing, verifying and providing feedback to get it the way I wanted. By the next Monday, I had the basis to a journaling application, AI feedback built into the tool and the start to my campaign to help others learn the basics of investment trading. All built on the VOS structure and all developed at 40x the speed it would have taken an engineer several months to do just a few years ago.

A product small enough that I could hold the whole thing in my head, but real enough that real users were going to use it. Before I knew it, I had built a fully functioning application, deployed on AWS infrastructure and a way for others to use it all by following our CID framework. Where code development is no longer the bottleneck, instead it has now shifted to the human intent and verification of what is coordinated and built by AI.

3. One shipped feature, start to finish

I'm going to walk through one specific VOS now, start to finish, because the chapter should have a worked example and I want you to see how realistic it is for any one person to follow this methodology.

The feature was called Trade Journal, the foundational VOS for creating the Trade Codex. This was the whole premise to support any trader and create a strong practice which is fundamental for anybody looking to trade, have a journal.

The WHY Section of the VOS was three sentences. *The core product: a trade journal where users log stock and options trades, track P&L, view history, and get AI-powered analysis. Supports multi-leg options (verticals, iron condors), scaling entries/exits, and calendar views. The journal is the foundation everything else builds on.*

The WHAT section was the acceptance contract. It was written in Gherkin, a language for writing specifications in a form that reads like English but is executable. The whole acceptance contract was about fourteen lines. I'm not going to reproduce all of it here because it's in Appendix B, but I want to show you one of the four scenarios so you can see what the sentence shape actually looks like:

```
Scenario: User logs a stock trade with scaled entry
  Given a user has an active trading account
  When the user logs a buy order for 100 shares of AAPL at $185.50
  And the user logs a second buy order for 100 shares of AAPL at $182.75
  Then the journal shall display both fills as part of one open position
  And the average entry price shall be calculated to the cent
  And the position size shall reflect the total share count
```

That is the kind of sentence I had to learn to write. Notice what it doesn't say. It doesn't tell the AI which database to use or what schema to write or what library to import. It says what has to be true when somebody logs two buy orders for the same stock at different prices. The AI handles the rest. The contract is the floor.

The gist of the rest of the contract was for every trade, it gets logged as a line item in the journal, we support both options and stocks, and we allow for users to scale in and out of trades while capturing the history of all logged trades. That was it, written as a Gherkin contract.

The HOW section was six lines and I barely wrote it. The AI orchestrator helped me write it and produce the tasks we expected to deliver. The HOW is the section the AI tool and Orchestrator own. The HOW said we were going create a form, have trade persistence, calendar view with P&L and trade detail page to capture the details. None of that was code. All of it was design decisions, made before any generation happened.

The CONTEXT section was three files. The journal module, database to support it, and the trade form to capture the details. That was it. I didn't include the entire codebase. I didn't include or direct the AI needs in another VOS, just what was needed for this VOS to deliver on what it needed with the context it needed. I didn't include anything not directly relevant to what the generated code needed to understand. This is important because we can get lost in trying to bring in everything and it is not needed. The evolution of AI has made understanding pulling in context simple and with that it makes generating and producing results that can actually be verified that much simpler.

The OUTCOME section was simple: *trades logged per active user per week, three trades a week per active user and trade count in database grouped by user and week. That was the goal and the outcome we strive to achieve to validate that this VOS drives value and is usable by a customer.*

As a result of this VOS, the system generated the code in about twenty minutes. I watched it happen. It set up the environment, created the database, set up the journal, gave me a local dev to see it after it was completed and created the tests. I ran the tests. They passed. I then ran the acceptance contract against the produced site, It also passed on the first try. I sat there amazed at what had just occurred, someone who does not know code, or sees code and just does not know what it does, just created my first VOS, had it developed through AI orchestration and now I had a site, that I built in front of me. A smile crept across my face in the middle of the night, just amazed at what just occurred. That was the beginning of what later transpired into an application that now has 20+ VOS created over four hundred thousand lines of code and is a tool to help anybody learn the basics of trading. All created by someone who still hasn't written a piece of code but had the idea to create something. It all starts with an idea, and now those ideas can become reality.

I want to be honest about something though, because the first VOS I just walked through is the kind of moment that makes you think this is going to be easy every time, and the truth is it isn't. Not every VOS shipped clean on the first generation. The second VOS I wrote, on Trade Codex, was the AI Trade Analysis feature, the part that takes a completed trade and gives you intelligent feedback through Bedrock. The first generation of that one was wrong. Not catastrophically wrong, but wrong in a way I had to learn to see.

The contract I had written said the analysis should "provide context-aware feedback on the trade, including pattern recognition and improvement suggestions." That sentence is the kind of sentence that would have been fine in a requirements doc I handed to an engineer ten years ago. The engineer would have asked me what I meant. The AI didn't ask. It generated a long block of generic trading-coach text that was the same for every trade. Pattern recognition that wasn't actually looking at patterns. Improvement suggestions that weren't tied to anything specific in the trade I'd just logged. It looked like analysis from a distance. Up close it was filler.

I sat with David on a video call the next morning and we went through what had happened. He didn't say I'd written a bad VOS. He asked what I was trying to actually have the AI do. We rewrote the contract together. The new contract had specific scenarios. Given a closed trade with a 5 percent loss and an entry that violated the user's stated risk-per-trade rule, the analysis shall name the rule violation and reference the specific entry. Given a closed trade with a 3 percent gain on a setup the user has logged seven losing trades on previously, the analysis shall name the pattern and suggest reviewing the setup. Five scenarios, each one tied to a specific behavior I wanted to see.

The regenerated code passed the new contract on the first run. The analysis on each trade actually said something true about that specific trade. And I learned something I would not have learned without the failure: the difference between a sentence I wanted an engineer to interpret and a sentence I wanted a machine to execute is the entire job. When the AI generated something generic the first time, it wasn't because the AI was bad at the job. It was because I had given it the kind of sentence that lets a human fill in the gaps. The machine doesn't fill in gaps. It does what the sentence says, and if the sentence is vague, the output is vague.

By the fifteenth VOS the magic had faded into rhythm. The first one had me staying up to watch the build complete because I had to see whether it would. The fifteenth one I wrote on a Tuesday morning, ran the generation in the background while I made coffee, came back and signed off on the harness output, and moved on to the next feature. That is what calibration looks like. The wonder doesn't disappear, but it stops being wonder and starts being work, and the work is the part that actually compounds. A pod that ships its first VOS in awe and its fifteenth VOS in routine has crossed a threshold the methodology was built for.

That's what shipping a VOS looks like. It looks like a Friday afternoon when you write down what you want, a Friday evening when the code gets generated and tested, and a completed feature before going to bed as it deploys to the cloud ready for use. It seems too good to be true, almost magical. Almost like it is too good to be true, however, this methodology is designed so nothing in it requires an expertise in software languages, because the bottleneck and table stakes to be able to create something amazing is at our fingertips today. And it is only the beginning of what "the art of the possible" really is.

4. Three things that surprised me

I want to share with you the three things that surprised me most about this journey. I feel this is important because the barrier to understand code and develop a working product has reduced tremendously with AI and I wanted to share my experiences.

The first thing that surprised me is how easy it was to get started. Yes, it is probably a lot to get started, to set up your tool, but once you get going, it really is just something that you interact with and as a result something great comes out the other side of it. You can speak to the tool like you would a human, type what you mean, be specific, and it will help shape what you want. The key here is holding true to what you want to see and don't be afraid to iterate on it and show it what you want and talk to the tool until you get it the way you want. You don't have to understand code, but you do have to have a vision and that vision will help you develop and be proud of what you develop with AI.

The second thing that surprised me is intent and clarity of that intent is important to get it right the first time. Machines and tools are getting better at interpreting meaning and what we want to produce, but it is not perfect. You learn a lot by reading and seeing what the tool is doing, as a result you quickly find out that the clearer you are with your intent and the better you write your Gherkin contracts, the better results you get.

The third thing that surprised me is this, why was I so afraid to do this in the first place!

I always looked at software development as a skillset needed, a language I needed to learn, or just the skills needed to get started was just too high for me. Glenn and David were always ready to support me and wanted to share with me how to develop something and to try it! I was always resistant, all through my career, but when I did try it, it opened a whole new world. David and Glenn tell me all the time that I sometimes create something that they feel is better or has a better look than what they can create. My artistic creations and how they got added into an application has surprised my peers that we quickly found out that you don't have to be an engineer to create something amazing. Just having the idea, a methodology to support it, you can create something quick and beautiful that fits your style. The hardest part is just getting started. So, my call to action for you is go try it, you may surprise yourself what you are able to create by just trying it. Don't be afraid of it, just try it and learn. You never know, it could open a whole new world for you that you never knew that existed, or that you were capable of doing this before.

5. What this means if you picked up this book without an engineering background

Here's the last thing I want to tell you.

I want to say who I think this chapter is for, because if you got this far you probably recognize yourself in some piece of it, and I want to name what I'm seeing. Three kinds of people I know are reading this book.

The first is the operations person who has spent a career being the one who explains what the customer needs to the engineers. You write requirements that engineers treat as usable. You sit in the meeting where the product manager and the lead developer disagree about what the feature should do, and you're the one who notices that they're talking past each other and reframes the question. You have been close to the work for ten or fifteen or twenty years and you have never written a line of production code. You also know things the engineers don't. You know what the customer steering committee said last quarter. You know why the renewal cycle peaks in March. You know which workflow change will land and which one will be ignored. The methodology in this book lets you keep doing that work, plus the next step, the one that used to require an engineer in the middle.

The second is the domain expert. The clinician who has been writing clinical-decision-support rules for fifteen years and watching engineers translate them into systems that almost work. The compliance officer who has been writing the requirements that get audited every quarter. The CFO who can describe the month-end-close in their sleep but has never built the tool that runs it. You have been the source of truth for software for a long time. You have watched the translation step lose half of what you said in the handoff. The methodology in this book lets you author directly, in language a machine can act on, with a verification gate you control the outcome of.

The third is the founder who has the idea and no engineering co-founder. You can describe the product in detail. You can articulate the customer who would use it. You have been told for the last five years that the first hire has to be a CTO because you can't ship software without one. The methodology in this book lets you reach product-market fit before you make that hire, from a position of strength, with the customer feedback already in hand.

I don't know which one of those three you are. I don't know if you're none of them, exactly, but you recognize the pattern in your own work. What I do know is that the methodology in this book widened the door for someone who had been outside it for two decades, and I am ordinary in the way most readers will be ordinary, which means the door is wider than the industry has admitted.

You don't need to be a software engineer, understand code, or have the technical background to do what I did. You need to learn three things. You need to learn to write a VOS much like you are creating an idea for a friend along with how you know it will work (dream big!)

Second, don't be afraid to use AI to ask questions, if you don't understand ask, If you see something and don't know what it means, ask. You can gain so much knowledge and experience by just watching and reading how AI creates your idea. You start to see trends, you start to see ways of working and when it comes up again, you know what is happening! That is the beauty of this approach is not only do get to create something you didn't know you could, but you also get a chance to learn. I often find myself asking my AI support agent, what do you recommend we do here, why is it this way, help me understand what makes sense here. All of this is helpful context that I can then feed into the VOS to help my AI support agent create something even better. All that to say, don't feel like you must stay in a box, ask anything and watch what happens.

Lastly, for the third item, have fun doing this. I have grown so much closer with Glenn and David who I consider close friends of mine galvanized on this journey we embarked on here at Alchemaize. Being each other's biggest cheerleaders as we crank out 35 apps in 100 days and creating content that ranges from a new framework (this book), business applications, games, health and wellness apps, mobile apps and so much more has been an incredible journey. There is so much joy out there and what I recommend is start with creating something for yourself, that you can use, that you can find passion in. Just by doing that you will find creating it is easier and more enjoyable because it is a passion of yours. That is the key, do it for you and you never know what will happen beyond that. But at the end of the day, you feel good knowing you were able to create something meaningful.

I want to share one more thing before I let you go. About four months into this, I was sitting at the dinner table with my wife and my son, and my wife asked what I had built that week. I told her, in plain language, what the application I had shipped on Tuesday actually did and who it was for. She listened the way she always does when I talk about work, and at the end she said, that is the first time you have described what you do for a living and it sounded like you made something. Twenty years of being good at my job and I had never said that sentence before. The thing I want you to know is that the methodology did not make me good at my job, because I had been good at my job for two decades. What the methodology did was let what I had been good at land in something I could point to. That is the quiet thing nobody warns you about when you start. The work is the same work it always was. The output is what changed, and the output is what your family and your customers and your colleagues can finally see.

Go try it. Start with one VOS, on something small you actually care about. Write down what you want, precisely enough that a machine couldn't misread it. Let it produce something. Let someone verify. Then do it again, on the next thing. That's the pipeline. The rest of the book is detail underneath it.

PART III

The Enterprise Layer

Chapters 9 through 11.

What changes when CID scales above the pod. Streams replace projects. Four numbers replace the dashboard. Compliance moves from quarterly audit to generation-time enforcement. The enterprise layer is ELCID, and it is the half of the book CTOs and VPs of Engineering will read first.

CHAPTER 9. STREAM FUNDING

I have spent most of two decades watching enterprises adopt Agile. I have sat in the rooms where the adoption starts, the rooms where the first quarter's results get explained, and the rooms, two years later, where somebody quietly cancels the transformation program without saying the word "cancel." The pattern is so regular that I have an internal shorthand for it. The ceremonies stick. The org chart reshuffles. The methodology gets certified, branded, measured, photographed at an offsite. And the money still flows the old way.

That last sentence is the whole chapter. If you only read one line of Part III, read it. A methodology adopted without a matching change to the funding instrument is a decoration on top of a budget. The budget wins. It always wins, because the budget is the thing the company can actually enforce, and the methodology is a set of suggestions the budget tolerates.

The frameworks at the team layer have been the subject of the first eight chapters of this book. The frameworks at the enterprise layer are the subject of the next three. But there is one instrument underneath both layers, and it has a specific shape, and that shape is incompatible with CID. It is project funding. The chapter names what project funding is, why it breaks under the new constraint, and what replaces it.

What replaces it is an intent stream. A persistent budget pointed at a persistent outcome, with a three-line structure (baseline, multiplier, sunset) and a monthly review cycle in which the portfolio of streams gets reshaped by the people holding the outcome metrics. This is not a new idea in its components. Lean budgeting has existed in some form for twenty years. What is new is that the methodology above the budget now makes the stream shape coherent, and the prior budget shape incoherent.

Enterprise CID is dead on arrival without stream funding. The Field Note after Section 5 is Casey's, because the honest view of how Alchemaize actually funded its streams during the 100-day sprint is one Casey should tell, not me.

1. The project-funding pathology

Every enterprise I have worked with funds software the same way. A VP or a business unit owner identifies a need. A project gets scoped: what it will deliver, by when, for how much. Finance approves the project, typically on an annual cycle with quarterly gates. Engineering staffs the project, typically with people pulled from other projects they were already staffing. The project runs for somewhere between nine and eighteen months, discovers halfway through that the scope was wrong, files a change request, waits three months for the change request to clear governance, and ships a compromise version at month eighteen. Everyone at the table knows this cycle. They planned for it. The eighteen-month project has nine months of actual work inside it, and the other nine are the tax the budget instrument extracts.

The pathology is structural, not accidental. A project is a promise: here is what you will get, by when, for this much. The promise is the thing finance can approve, and to be auditable it has to name scope, dates, and a dollar figure, all settled at the top of the cycle, before the team doing the work has learned enough to say any of them with confidence.

That is the false precision the project instrument demands. The team is asked, at month zero, to commit to a shape of work it cannot yet see. It commits anyway, because the budget cycle requires it and the people approving the budget have learned to treat any answer other than a commitment as a dodge. Six months later the team discovers the commitment was wrong in specific ways and files a change request. The change request is expensive to process, not because the processing is hard, but because every downstream plan, the calendar, capacity planning, adjacent projects, was keyed off the original commitment. Changing one project's scope is a ripple that takes a quarter to settle.

That quarter of settling is the process cost of the original false precision. Multiplied across a portfolio of forty projects in an enterprise, it is a substantial fraction of the engineering organization's time, and it is purely metabolic. The cost the organization pays for having demanded scope commitments at a moment when those commitments were not yet knowable, with nothing shipping out the other side.

Under the old constraint, this pathology was tolerable. Humans were the bottleneck. The eighteen-month cycle was, roughly, the rate at which a human engineering organization could actually learn, replan, and deliver. The budget cycle and the learning cycle were not perfectly matched, but they were close enough that the friction was manageable. An enterprise building software by typing could afford to spend nine of every eighteen months on the tax, because the other nine were fully utilized by the typing itself.

The new constraint will not tolerate this. A CID pod ships continuously. It does not batch work into a sprint and the sprint into a program increment and the increment into a project. It ships VOSs, verified, into production, at a cadence measured in days rather than months. If the budget cycle is still running at eighteen-month resolution, the pod is producing learning six times faster than the instrument measuring its funding can react. The pod will out-learn its own budget, repeatedly, within the budget's first quarter.

What happens then? The pod discovers the scope was wrong, but in a different way than the old cycle discovered it. The pod discovers the scope was wrong in week three, not month six. And the pod discovers it continuously, because the pipeline review surfaces the learning every week. The pod cannot file a change request every time the scope shifts. The cycle would consume the pod's calendar. Instead, the pod does what a CID pod does: it adjusts the next VOS. The adjustment is silent. It does not surface to the budget instrument at all. The budget, back at the enterprise layer, is still holding the original scope commitment, while the pod is three scope changes deep into work the budget has never seen.

This is the moment where project funding becomes incoherent. The budget is tracking one thing. The work is tracking another. The two have decoupled, not because the pod is going rogue, but because the pod is doing what the methodology asked it to do, which is to let continuous learning reshape the work continuously. The budget was never designed to absorb continuous learning. It was designed to hold a fixed commitment and reconcile at the end.

In practice, in every CID engagement I have seen, the decoupling gets papered over in the same way. Someone in the PMO writes a quarterly variance report. The variance report explains, in careful language, that the team has shipped outcomes roughly aligned with the original scope, plus or minus some differences it describes as refinements. The variance report is believed, because the outcomes did roughly ship. The budget survives another quarter. Nobody looks too closely at the gap between what was committed and what actually got built, because looking closely would require admitting that the commitment was never real in the first place.

That is the pathology. Project funding asks for a precision the work cannot provide. The work provides it anyway, because the budget requires it. And the gap between the commitment and the reality becomes a quiet fiction the organization maintains to keep the budget instrument functional. Under the old constraint the fiction was small enough to ignore. Under the new constraint the fiction widens every week.

Eighteen months of fiction is what we are replacing.

2. The intent stream

An intent stream is a persistent funding commitment pointed at a persistent outcome. That is the whole definition. Everything else is the shape.

The shape has three lines. A baseline allocation that funds the pod and the platform it runs on, sized to keep the pod working at a steady rhythm for one quarter. An outcome multiplier that adds funding if the stream's primary KPI moves by a specified amount within a specified window. And a sunset trigger that terminates the funding if the KPI fails to move by a smaller specified amount within a longer specified window. Three lines. The whole funding instrument fits on a single index card.

The baseline is the boring line, and it does most of the work. A CID pod is three people plus the model-inference and compute costs that run its generation and verification stages. At Alchemaize in early 2026, a single pod's baseline ran about $180,000 per quarter, counting salaries at market, platform consumption at observed rates, and a small tooling overhead. In a larger enterprise with higher salary bands and heavier compliance overhead, the baseline runs higher. I have seen it sized between $250,000 and $400,000 per pod per quarter in financial services. The number is less important than the structure. The baseline is a steady commitment, known at the start of the quarter, paid in a predictable cadence, that funds the pod to keep doing the work.

The multiplier is the interesting line. It is a pre-agreed additional amount of funding, released when the stream demonstrates that its work is moving the outcome it is pointed at. The mechanism is simple. The stream has a primary KPI, stated in the charter, with a baseline and a target. The multiplier releases when the KPI moves by a specified fraction of the gap between baseline and target within a specified window. At Alchemaize we used thresholds like 20 percent of the gap within a quarter, with a 50 percent multiplier (meaning, if the KPI moved that much, the stream got an additional half of its baseline). The exact numbers are organizational choices.

The structure is invariant. Moving the outcome earns more funding. Not shipping more features. Not passing more sprints. Moving the outcome.

The sunset line is the line a CFO will read first. It is the exit mechanism. If the KPI fails to move by a smaller specified amount within a longer specified window, the stream is flagged for sunset at the next Monthly Intent Review. The flag does not mean the stream is terminated immediately. It means the stream has entered a defined thirty-day wind-down in which it either articulates a concrete reason the KPI will move in the next window, or transitions its pod to another stream. The typical numbers I have used are 10 percent of the gap within two quarters as the sunset threshold. If a stream cannot move its primary KPI 10 percent of the way to target in six months of continuous work, the stream is wrong about either its hypothesis, its pod, or its scope. Any of those is a reason to reallocate the funding.

Three lines. Baseline, multiplier, sunset. A five-hundred-word stream charter carries all three, plus the KPI definitions, plus the compliance extensions in effect, plus the pod composition. The charter is one page. It is the operating document of the stream for the quarter. A CFO can read it in under ten minutes.

Notice what the stream does not have. It does not have a scope. It has an outcome. The outcome is the KPI movement the stream is pointed at, stated quantitatively, with a baseline and a target. The scope, which is the specific set of features or changes the stream will ship, is a decision made continuously at the VOS cadence, inside the pod's Pipeline Review. The stream's funding does not depend on the scope. The scope is the variable the pod adjusts weekly in pursuit of the outcome the stream owns.

This is the inversion. Under project funding, scope is fixed and outcome is hoped for. The project commits to shipping features A, B, and C, and hopes the features produce the business result. Under stream funding, outcome is committed and scope is adjusted. The stream commits to moving a KPI by a specified amount, and ships whatever sequence of VOSs actually produces the movement. If the first three VOSs do not move the KPI, the next three are different VOSs, chosen by the pod based on what the first three revealed. The stream does not file a change request when the scope shifts. The scope shift is the work.

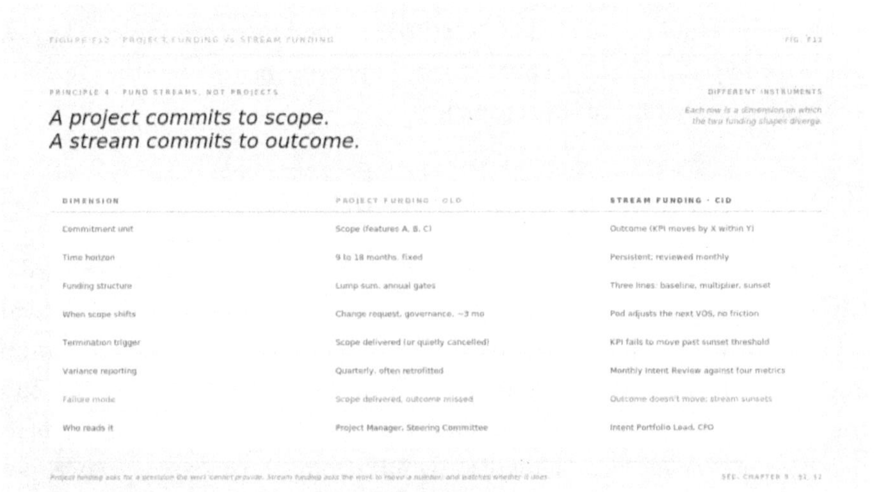

FIGURE F12 · PROJECT FUNDING vs STREAM FUNDING

PRINCIPLE 4 · FUND STREAMS, NOT PROJECTS

A project commits to scope.
A stream commits to outcome.

DIFFERENT INSTRUMENTS

Each row is a dimension on which the two funding shapes diverge.

DIMENSION	PROJECT FUNDING · OLD	STREAM FUNDING · CID
Commitment unit	Scope (features A, B, C)	Outcome (KPI moves by X within Y)
Time horizon	9 to 18 months, fixed	Persistent; reviewed monthly
Funding structure	Lump sum, annual gates	Three lines: baseline, multiplier, sunset
When scope shifts	Change request, governance, ~3 mo	Pod adjusts the next VOS, no friction
Termination trigger	Scope delivered (or quietly cancelled)	KPI fails to move past sunset threshold
Variance reporting	Quarterly, often retrofitted	Monthly Intent Review against four metrics
Failure mode	Scope delivered, outcome missed	Outcome doesn't move; stream sunsets
Who reads it	Project Manager, Steering Committee	Intent Portfolio Lead, CFO

Figure F12 · Project vs stream funding. *Comparison of project-based and stream-based funding patterns.*

Shipping in smaller batches is not a preference here; it is what lets the feedback channel function. Nicole Forsgren, Jez Humble, and Gene Kim's *Accelerate* documents smaller batch sizes as one of the technical capabilities that correlates with higher software delivery performance [1]. A stream that adjusts scope every week is a stream shipping in the smallest batches the organization can currently verify. The funding instrument has to match that cadence, or it fights the pod.

The stream also does not have a deadline in the project sense. It has a sunset window. These are not the same thing. A deadline says: if you have not shipped by this date, you have failed. A sunset window says: if the outcome has not moved by this date, the funding stops. The difference is the subject of the clause. A deadline is about delivery. A sunset is about effect. One of them is a scope instrument. The other is an outcome instrument.

This is a compact definition, and it is worth sitting with before the chapter moves on. The stream is a persistent pod, running a continuous pipeline of VOSes, pointed at an outcome, funded on three lines, reviewed monthly, adjusted quarterly, terminated when the outcome will not move. The pod inside the stream does CID. The enterprise around the stream does ELCID. The stream is the connective shape between them, and the funding instrument is what makes the shape hold.

3. The CFO objection and its answer

At this point in every CFO conversation I have had about this model, the CFO asks a specific question, in roughly the same words. How do I defund a stream that is not moving its KPI? The question is sometimes dressed up in process language, but the underlying concern is always the same. The CFO has a board. The board has fiduciary responsibility. A funding instrument that does not have a clear defunding path is, from a board's perspective, not a funding instrument. It is a subscription.

This is a fair question, and the answer has to be concrete enough to defend on a board memo. It is not enough to say that streams get reviewed and sunsetted when they underperform. The mechanism has to be legible, specific, and boring. Boring is the target state. A CFO who can read the mechanism and think "this is how we do budget reviews already, with different nouns" is a CFO who will back the model. A CFO who reads the mechanism and thinks "this sounds like a way for engineering to keep the money flowing indefinitely" is a CFO who will kill it. The difference is not the model. It is the clarity of the exit.

The exit is the sunset mechanism, and it has four parts.

The first part is the baseline window. When a stream is funded, the charter states a baseline window during which the stream is not eligible for sunset review, typically one quarter. This is the period in which the pod is getting its pipeline running, shipping its first VOSs, establishing its verification patterns, and beginning to instrument the outcome KPI against real production data. In that first quarter the KPI should be measurable but not yet meaningfully moving. Protecting the first quarter from sunset pressure lets the pod do the work that makes the subsequent quarters legible. Without this window, every stream is judged on its first eight weeks of shipping, which is the noisiest eight weeks of its life.

The second part is the KPI threshold. After the baseline window, the stream is expected to show KPI movement at a specified rate. The threshold is stated in the charter. I use 20 percent of the baseline-to-target gap per quarter as a default, scaled to the KPI's expected responsiveness. A latency reduction KPI may respond faster. A customer retention KPI may respond slower. The charter names the threshold at the time of funding, not at the time of review. This matters. The threshold is a commitment from both sides: the stream commits to moving the KPI by that amount, and finance commits to funding the stream at baseline while the stream is moving it.

The third part is the sunset review. Once per quarter, at the Monthly Intent Review, every stream that is under its threshold is flagged. The flag is not a sunset. It is a scheduled conversation. The stream lead has thirty days to return to the next review with one of three answers: a concrete plan for the KPI to move in the next window, a recommendation to reallocate the pod to a different stream, or a recommendation to sunset the stream outright. If the plan is credible, the stream continues. If the reallocation is better, the pod moves. If sunset is the right answer, the stream enters wind-down.

The fourth part is the wind-down itself. A sunsetting stream has thirty days to finish any VOS in flight, document what was learned, hand off any production systems to a receiving stream, and release its pod. The pod does not get disbanded. The pod gets reallocated. In practice a pod that has been doing CID together for two quarters has a specific operational literacy that the organization should preserve. The pod is the asset. The stream was the assignment. The assignment ends. The asset redeploys.

Those four parts are the mechanism. A CFO reading it sees what they need to see: a window, a threshold, a scheduled review, and a defined exit. The funding is bounded at both ends by rules stated in advance. I have sat in the rooms where boards approved this, and once the mechanism is written down the conversation is shorter than the first time through, because the mechanism looks like the other budget reviews the board already approves. Different nouns, same instrument shape.

What the mechanism does not do, and this is worth saying clearly, is absolve the organization of judgment. The threshold is a default, not a verdict. A stream at 18 percent of its threshold in a category where the expected movement is slow may be doing fine work; a stream at 25 percent in a fast-moving category may be underperforming. The Intent Portfolio Lead brings that judgment to the Monthly Intent Review, and the review makes the call. A board that is comfortable with the distinction between scaffold and judgment can approve the model. A board that expects the mechanism to do the judgment for them will eventually be disappointed by any funding model, project or stream. That is a governance problem, not a funding problem.

The sunset mechanism is not actually the thing that makes stream funding work. It is the thing that makes stream funding defensible. What makes it work is that the outcome multiplier gives the organization a feedback channel tight enough to see the KPI move within a quarter. Under project funding, the spend-to-outcome cycle runs at the length of the project, typically a year or more. Under stream funding it runs at the length of the KPI movement window, a quarter. A feedback channel that updates four times a year is a different instrument from one that updates once. Most of what the CFO gains comes from that compression, not from the sunset mechanism. The sunset is what the board reads; the multiplier is what the business feels.

A funding model a CFO will defend has a clear exit and a short feedback channel. Stream funding has both.

4. The Portfolio of Intents

The stream is the unit. The portfolio is the shape of many streams together.

A Portfolio of Intents is the curated, prioritized set of outcomes an organization is funding at any given time. Each intent has a stream pointed at it. Each stream has a pod inside it. Each pod runs a pipeline. The portfolio is the document that names all of the outcomes the enterprise is pursuing, tracks the four portfolio metrics rolled up across the streams, and flags the streams whose funding is due for reallocation, multiplier, or sunset.

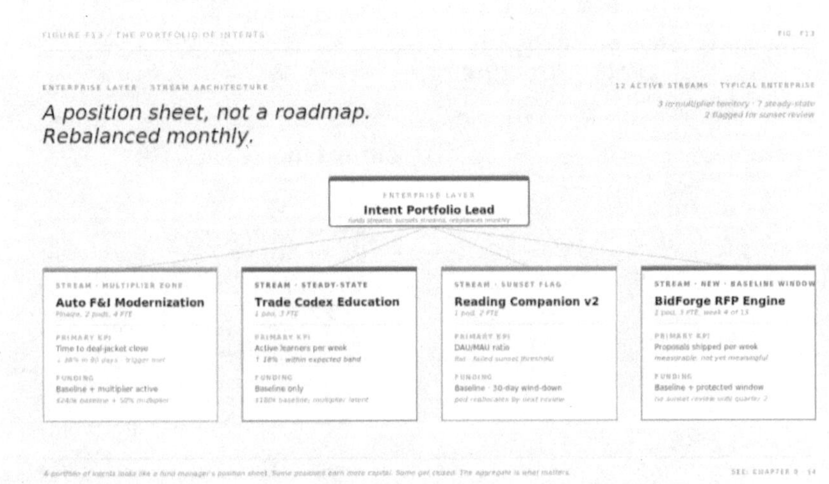

Figure F13 · Portfolio of Intents. *Intent-portfolio structure at the enterprise layer.*

In traditional program management, the equivalent document is a roadmap. A roadmap lays out features against dates, grouped by product or initiative, usually spanning twelve to eighteen months. The roadmap is the artifact that gets presented to the board, referenced in earnings calls, and used internally to coordinate engineering against sales, marketing, and customer success. It is, in most of the enterprises I have worked with, the single most consulted document in the organization. It is also, in most of those enterprises, significantly wrong by month three.

Scaled Agile's SAFe framework formalizes this shape at the portfolio layer with epics, value-stream budgets, and Lean Portfolio Management (`scaledagile.com`) [2]. The epics get written, the value streams get drawn on a poster, the Lean Portfolio team meets on a cadence. The funding instrument underneath is still a project budget in most SAFe implementations I have seen, and the epics behave like projects with different vocabulary. A Portfolio of Intents is not an epic backlog with a new cover sheet. It is the swap of the underlying instrument.

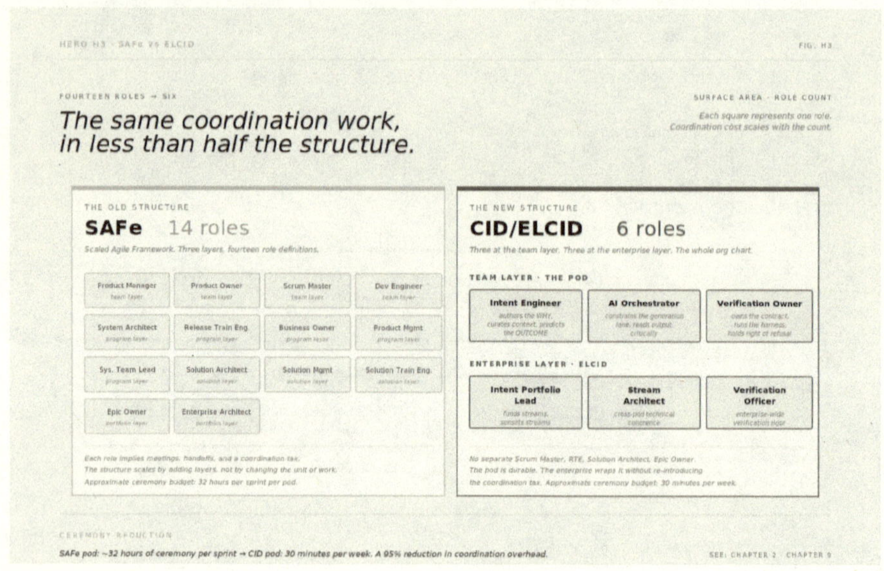

Side-by-side comparison of SAFe scaling vs ELCID stream funding.

A Portfolio of Intents is not a roadmap. It is closer, structurally, to a fund manager's position sheet. Each intent is a position. Each stream is the capital allocated to the position. The outcome metrics are the returns. The portfolio is rebalanced monthly, the way a fund rebalances monthly. Positions are opened when new intents are funded. Positions are closed when intents reach their targets or hit their sunset thresholds. Positions are sized up when the multiplier releases. Positions are sized down when the pod is partially reallocated. The Intent Portfolio Lead, in this analogy, is the fund manager.

This is not a loose metaphor. It has operational consequences. A fund manager does not defend every position. A fund manager is judged on portfolio return, not position return. Some positions will underperform. Some will be closed. The discipline is not in holding every position but in the ruthlessness of the rebalance. The same discipline applies to a Portfolio of Intents. The Intent Portfolio Lead is not defending every stream. The Lead is optimizing the portfolio's aggregate outcome, which requires closing some streams to feed others.

In practice, a healthy portfolio of intents at an enterprise I worked with in late 2025 ran something like this: twelve active streams, three in outcome-multiplier territory (moving their KPIs faster than expected), two in sunset-flag territory (moving their KPIs slower than expected), and seven in steady-state (moving at roughly expected rates). The two in sunset flag were the month's agenda items. One was reallocated, the other was sunsetted. Two new streams were opened into the freed funding. The portfolio at month-end had twelve active streams again, but not the same twelve. That is the rhythm. The aggregate looks steady. The constituents churn.

The four metrics at the portfolio level are the same four metrics the pods track, rolled up. Aggregate intent cycle time (how fast VOSs are moving through the pipeline across the portfolio). Aggregate first-pass verification rate (what percentage of VOSs pass their contracts on the first attempt). Aggregate outcome KPI delta (the weighted movement of the portfolio's outcome metrics). And portfolio cost per shipped intent (total spend divided by VOSs shipped). These four numbers are the Intent Portfolio Lead's dashboard. They are the only numbers the Lead needs to run the review. Everything else is inside the stream charters, and the Lead reads charters when decisions require it, not on a schedule.

The Portfolio of Intents does something a roadmap cannot do, which is keep its promises. The promise of a roadmap is that the listed features will ship on the listed dates. The promise is broken by month three, every time, and the organization manages around the broken promise with ad-hoc replanning. The promise of a Portfolio of Intents is different. It does not promise features. It promises outcomes. And the outcomes, unlike features, are stated in terms the business can actually measure and the streams can actually optimize against. When an outcome does not move, the stream gets flagged, and the portfolio rebalances. The promise the portfolio makes is that unmoved outcomes will be named, addressed, and either corrected or closed. That promise can be kept. A roadmap's promise cannot.

The portfolio is the artifact the board sees. It replaces the roadmap in every review meeting, every quarterly business review, every earnings-preparation cycle. It is shorter, more honest, and more accurate than a roadmap, because it lists outcomes rather than features, its currency is KPI movement rather than delivery commitment, and the list gets rebalanced monthly rather than annually. A board that has learned to read it will not want a roadmap back.

The portfolio composes. The roadmap does not.

5. The Intent Portfolio Lead

The role that manages the portfolio is the Intent Portfolio Lead, and it is a new role, not a renamed VP of Product, head of the PMO, or Chief of Staff. It is the consolidation of the functions those three roles were performing under the old constraint, performed now by a single person operating against a different instrument.

A VP of Product, in the old shape, owned the roadmap. The roadmap was a coordination document, not a capital-allocation document. The VP of Product argued for features, prioritized features, defended features, reshuffled features. Funding came from elsewhere (Finance, typically through project budgets approved annually). The VP of Product did not allocate capital. The VP of Product curated scope inside a capital envelope somebody else set.

A head of the PMO, in the old shape, owned the process. The PMO ran the planning cycles, tracked the status, produced the variance reports, and maintained the schedule. The PMO was the audit layer on top of the project portfolio. Like the VP of Product, the PMO did not allocate capital. It tracked how capital already allocated was being spent.

The Intent Portfolio Lead allocates capital. That is the first thing to say about the role. The Lead decides which streams get funded, at what baseline, with what multipliers, against what thresholds. The Lead decides which streams get closed. The Lead decides which pods get reallocated to which streams. The Lead reports, in most enterprises I have worked with, to the CFO or the CEO, not to the Chief Product Officer and not to Engineering. The reporting line matters. It is the reporting line of a person with a fiduciary relationship to the business, not a person with a delivery relationship.

The Lead's primary artifact is the Portfolio of Intents. The Lead maintains it, updates it monthly, presents it at the Monthly Intent Review, and answers for it at the quarterly board review. The Lead's primary meeting is the Monthly Intent Review itself, ninety minutes, once a month, with the stream architects, the stream leads, and the Verification Officer. That meeting is where the portfolio gets rebalanced. It is the only meeting the Lead runs. Everything else is asynchronous.

What the Lead's calendar actually looks like is instructive. It has approximately one recurring meeting, the Monthly Intent Review. It has approximately three to five ad-hoc conversations per week with stream leads, usually triggered by a specific decision that is ready to be made (a threshold reset, a reallocation, a new intent being scoped). It has approximately one hour per week of portfolio analysis, reading the four metrics and looking at which streams are trending. It has approximately two to four hours per week of writing, producing the monthly portfolio narrative and the occasional board memo. Total load, roughly, is ten to fifteen hours per week. The Intent Portfolio Lead role, done correctly, is not a forty-hour-a-week operational job. It is a twelve-hour-a-week decision-making job, and the other twenty-eight hours are available for the Lead to do something else of value to the business.

This is a statement that will raise eyebrows in every enterprise that currently runs a PMO. The PMO is large. It has a head, a staff, a set of practices, a budget. The Intent Portfolio Lead replaces it at a load of twelve hours per week. What happened to the rest? Most of it was the tax of coordinating under project funding. The status meetings, the variance reports, the change-request processing, the schedule reconciliation, the cross-project dependency management. Under stream funding, those activities do not exist at the same scale. Streams do not have fixed scopes, so there is no variance between committed scope and delivered scope to reconcile. Pods manage their own dependencies at the Pipeline Review cadence, because dependencies surface inside the pipeline. The cross-stream coordination is handled by the Stream Architects (a separate role, not the Portfolio Lead) at a cadence that does not require a standing organization to support.

The Intent Portfolio Lead fails when the Lead starts managing streams instead of managing the portfolio. Every enterprise I have worked with produces this failure mode at some point in the first year. A stream has a problem. The Lead sees the problem, has an opinion about the problem, and steps into the stream to solve it. The first time this happens, nobody notices. The tenth time, the pod has stopped owning its work, and the Lead has acquired a part-time operational role inside the stream. The Lead's calendar fills. The portfolio view drops. The portfolio rebalancing gets sloppier, because the Lead is too deep in individual streams to see the aggregate. This is the failure mode. The Lead's job is to hold the portfolio. The stream architects and pod leads hold the streams. When the Lead reaches into a stream, the instrument breaks.

The mechanical check for this failure is simple. Look at the Lead's calendar. If there are more than three stream-specific meetings on the calendar in a given week, the Lead is probably managing streams. If there are fewer than three stream-specific meetings on the calendar in a given month, the Lead is probably not engaging with the streams enough. The middle is the target state, and the middle is narrower than it sounds.

The role itself is recruitable. In every engagement I have done, the Intent Portfolio Lead ends up being someone already in the organization: typically a senior product leader with finance fluency, a PMO head with product instincts, or a former CFO staff member with deep product exposure. The role does not require prior CID experience. It requires the ability to read the four metrics, the judgment to make the rebalance calls, and the institutional weight to stand behind them. In the Alchemaize context, that was David himself, during the 100-day sprint. At a larger enterprise it will be someone with fifteen to twenty years of mixed product and finance experience and no prior identification as an Agile coach. The role does not recruit from the Agile consultant pool. It recruits from the product-finance intersection, which is a thinner layer, and one of the supply constraints on ELCID adoption at scale.

The Lead is the role that makes the portfolio real. Without the Lead, the stream charters are documents in a shared drive and the Monthly Intent Review is a meeting somebody schedules. With the Lead, the portfolio is an instrument that allocates capital against outcomes, reviewed in a rhythm the organization can sustain, defensible to a board, and accountable for the four numbers the business actually runs on.

Field Note from the Test Pilot

Casey Robinson, Intent Engineer

How we funded the BidForge stream the week before we shipped it, and why a CFO reading this should be nervous

Let me tell you what actually happened when we funded BidForge.

David and Glenn and I were on a Tuesday afternoon video call. Glenn had put together the idea for a proposal response tool we were calling BidForge, a platform that helps companies decompose incoming RFPs and then recompose the response using AI, pulling in organizational strengths, identifying gaps, and generating a more automated, structured proposal. He'd shared the initial concept in GitHub and walked us through what the tool would do. I asked three questions about the workflow. How proposals get ingested. How the AI identifies what the org is strong on versus where the gaps are. How the final response gets assembled. David nodded, and by the end of the hour we'd agreed we would start building it. We had AI summarize the conversation so we had notes of what was discussed. That was the funding decision. Nobody wrote a charter. Nobody ran a threshold. Nobody set a sunset window. We said yes, and then

Glenn and I split up the first four VOSes between us that evening.

A week later we shipped the first working version, which still sort of amazes me to write.

I'm telling you this because the chapter above is honest about what stream funding is supposed to look like, and I want to be honest about what it actually looked like when we were the ones doing it. At Alchemaize in early 2026, "funding a stream" meant David and Glenn and I agreeing to work on something. We don't have a CFO, we don't have a board meeting, and we don't have a finance function. I'm the COO and the person closest to being a finance function, and my role in the BidForge decision was to ask whether the product was scoped narrowly enough that we could get a first version up in a week. That was the check. The check worked for us because we're three people who'd already had the conversation, more times than I can count, about what we were and weren't going to pick up.

At enterprise scale this is harder than the chapter admits, and I'd like to be honest about that too, because I've watched a couple of first-time implementations struggle right here.

The mechanism David describes is real. The four parts of the sunset work. The three lines of the stream charter work. The Monthly Intent Review works. We've run all of it with customers, and it does what it says it does. What I want to say, as the COO, is that the first time you try to run this inside a company that has a CFO and a board and an audit committee, you are going to have a conversation that is much harder than reading the chapter makes it sound.

The hard part is not the mechanism. The hard part is the judgment the mechanism asks for. A project budget is defensible because it is a commitment. You said you were going to deliver features A, B, and C, and you either did or you didn't. A stream budget is defensible only if the people in the room agree on what "moving the KPI" actually means for this specific business. If the KPI is clean (revenue, retention, conversion), the conversation is tractable. If the KPI is soft (engagement, satisfaction, some internal metric the business hasn't instrumented well), the conversation is rough. The sunset decision comes down to a judgment call, and the judgment call is made by people who have skin in the game. The first few times they make it, they will be wrong in both directions. They'll sunset a stream that was about to break through. They'll fund a stream longer than it deserved. That's what the learning curve on this looks like.

I've seen early engagements where the Intent Portfolio Lead spent the first quarter effectively recalibrating the thresholds, because the numbers set in the charter turned out to be either too tight or too loose. That's not a failure. It's the instrument being tuned. But it's uncomfortable, and it will read to a first-time board as the Lead not knowing what they're doing. The Lead needs air cover for the first two quarters of the model running. Somebody above the Lead, the CFO or the CEO, has to be willing to say out loud that the thresholds will shift while we learn the business's actual response curves. Without that cover, the Lead gets blamed for the instrument's first pass being noisy, and the model dies in its first review cycle.

The other thing I want to name is that the sunset decision is emotionally harder than any comparable project-cancellation decision in the old shape. A project cancellation happens because the project missed its dates. It's about delivery failure, which is a known category. A stream sunset is a judgment that the outcome the business chose to pursue is not going to be moved by the team pursuing it, in the window they were given. That's a different kind of decision. It names the outcome, not the team. And the people working on the stream have usually been working hard, shipping, iterating, doing the work the methodology asked them to do. The sunset is not a judgment about their effort. It is a judgment about the bet. But it will feel, to them, like a judgment about their effort, especially the first few times the organization does it.

An honest first implementation takes that seriously. It names the sunset decision in advance so everybody knows what the bet is and what the exit looks like. It does the sunset review in person, or on video with the full stream lead in the room, not in an email. It reallocates the pod, visibly, to a new stream that the organization is funding, rather than treating the sunset as a layoff trigger. And it documents, in writing, what the organization learned from the stream, so the next stream can avoid the same mistake. Those four things are not in the chapter, and they should be, because without them the mechanism hardens into something worse than project cancellation: a regular public firing of ideas that attached themselves to people.

So, for the CFO reading this book.

The model is defensible. The numbers work. The boards I've sat across the table from have approved it when the mechanism was shown to them clearly. What I want you to hear is that the model makes three demands on the organization that project funding did not. It asks the Intent Portfolio Lead to have genuine judgment about outcome movement. It asks the sponsoring executive to give the Lead air cover for the first two quarters of calibration. And it asks the organization to run sunset conversations with the specific care that keeps the pod reallocating rather than disbanding.

If you are not willing to commit to all three of those, the model will not work. It will look like it is working for one quarter, and then it will collapse into the first cancellation conversation, and the organization will decide that stream funding is unworkable, when what was actually unworkable was the organization's appetite for running the mechanism with the care the mechanism requires.

We got away with the napkin-sketch approach at Alchemaize because we are three founders in three states running a company that ships thirty-five applications in a hundred days, and the trust between the three of us is the instrument that lets the napkin sketch work. At enterprise scale the trust has to be built by the

> mechanism rather than assumed by the room, and that is more work than the chapter admits. It's still worth doing. I just want the CFO reading this to know what they're committing to before they commit.

6. The transition

Stream funding is one of three enterprise-layer changes CID requires to work at scale. The other two are metrics and compliance, and they are the subjects of Chapters 10 and 11. A portfolio of streams, funded on three lines and reviewed monthly, is the financial instrument. A set of four numbers, visible on one dashboard, is the measurement instrument. A set of modular extensions, opted into by stream, is the compliance instrument. The three together are ELCID. Without all three, the methodology stalls at the team layer and never composes into an enterprise practice.

The funding instrument is the one most enterprises will touch first, because it is the one most visible to finance, and finance is the function that will either bless or kill the adoption. This chapter is the argument to bring into that meeting. The next chapter puts the numbers on the portfolio that a CIO can read in thirty seconds.

Money flows where work flows. Make the money match the work.

Notes

1. Nicole Forsgren, Jez Humble, and Gene Kim. *Accelerate: The Science of Lean Software and DevOps*. IT Revolution, 2018.
2. Scaled Agile, Inc. *SAFe 6.0 for Lean Enterprises*. Scaled Agile, 2023. scaledagile.com/safe-big-picture (accessed 2026-04-22).

CHAPTER 10. FOUR NUMBERS, ONE MEETING

I've spent twenty-five years watching enterprise engineering dashboards grow. Every one of them told a story. Almost none of them told the story the CEO was actually asking about.

The dashboards I mean are familiar. Velocity by team, rolled up by program, rolled up by portfolio. Burndown by sprint. Predictability percentage by PI. Capacity utilization by resource pool. Feature completion by quarter. Release throughput by train. Change fail rate, deployment frequency, lead time, mean time to recovery. The last four are the DORA keys, surfaced by the research Nicole Forsgren, Jez Humble, and Gene Kim reported in *Accelerate* and carried forward by the State of DevOps surveys [1]. Some of them genuinely useful in a tight engineering context. Most of them decorative once they reached the executive floor. All of them, taken together, a kind of pointillism. You could stand close and see a lot of numbers. You had to stand very far back to see a picture, and when you did, the picture wasn't always the one the numbers were supposed to be telling.

The board at my last enterprise engagement asked a simple question at each quarterly review. *Is the engineering investment producing business results?* The dashboard in front of them had forty-two charts on it. None of them answered.

That's the gap this chapter is about.

1. The graveyard

The methodology we inherited from twenty-five years of Scrum and SAFe came with a metrics library. Every framework does. The library was reasonable for the constraint the framework was designed against, and, like the framework, it doesn't survive the move to a generation-dominated pipeline. The cost of carrying it forward is not neutral. The old metrics actively mislead under the new constraint. They give the appearance of signal where there is only activity, and the organization responds to what the dashboard is showing, because that is what dashboards are for.

Before we put four numbers on the table, we have to bury the ones those four are replacing. I'm not going to spend a page on each. I'm going to name them, name what they were measuring, and name why they no longer answer any question worth asking.

Velocity. What it measured was typing throughput, dressed up as value throughput. Story points completed per sprint, rolled up across teams, charted against a historical average. For twenty years this was the number an engineering manager could show a director to justify the quarter. Under a generation-dominated pipeline, velocity measures what the AI generation layer produced, which is the cheapest step in the system. Congratulations to the team that ships a hundred story points of wrong software in a sprint.

Story points. A coordination instrument, never a measurement instrument. The point was to let humans agree on the relative size of work items without committing to a unit of time, so that estimation could happen without the psychic weight of promising a deadline. Story points existed because human typing time was variable, uncertain, and politically charged.

The generation step is now not the slow step, and the slow steps that remain (intent clarification, verification) resist story-point estimation in the same way that "how long does it take to write a good specification" has always resisted it. Story points survived because they were useful to Scrum. Scrum is not the framework we are running. Story points are an artifact of a ceremony that no longer runs.

Burndown. A sprint-artifact. It visualized the consumption of a fixed scope against a fixed time window. Both the fixity of the scope and the fixity of the time window belonged to a framework that organized work in two-week boxes because that was how long a human team could hold context. Continuous flow does not produce a burndown curve. Continuous flow produces a trend line on four different numbers, which is what the rest of this chapter is about.

Capacity utilization. This one is the tell. Capacity utilization measures whether the humans are busy. It was useful when the humans were the bottleneck, because underutilized humans were an expensive form of slack. It is actively harmful now, because the humans are no longer the bottleneck and measuring their utilization re-centers the metric around the wrong thing. A CID pod operating at 60 percent human utilization and shipping four VOSes a week is beating the pod operating at 95 percent utilization and shipping one. The dashboard that penalizes the first pod is lying.

Predictability. The number that SAFe made into a virtue. The percentage of committed features delivered per PI, averaged across trains, celebrated at program reviews. Predictability is a virtue only when you are confident the right work has been committed to. If the intent portfolio is pointed at the wrong outcomes, predictability tells you how reliably you are failing. There is a version of this industry in which high predictability against a flat business outcome was held up as success for five consecutive years. I've been in those rooms. I have written the slide that put a green check mark next to a predictability number in a quarter where the product KPI moved zero.

That's the graveyard. Five metrics, five funerals, one paragraph each, all of them honorable service records for the constraint they were designed against, and all of them retired.

I notice I've used the word *dashboard* more times on this page than I will in the rest of the chapter. Dashboard is a 2010-vintage word for the same thing enterprises have always wanted: a single pane of glass that tells the CIO whether the thing is working. The word isn't the problem. The content is. What you put on the single pane of glass determines whether the question gets answered or decorated.

2. The four

A CID pod tracks four numbers per week. Same four numbers at every level of aggregation. The pod's Pipeline Review uses them. The portfolio's monthly review uses them. The CIO's thirty-second read of the engineering organization uses them. Same four.

The four are:

1. **Intent cycle time.** Median hours from a VOS entering the DRAFTED state to the same VOS reaching SHIPPED-AND-OBSERVED.
2. **First-pass verification rate.** Percentage of VOSes that pass the acceptance contract on their first generation run, with no rework.
3. **Outcome KPI delta.** Movement in the business metric the stream is pointed at, measured against its baseline.
4. **Cost per shipped intent.** Fully loaded stream spend divided by the count of VOSes shipped in the period.

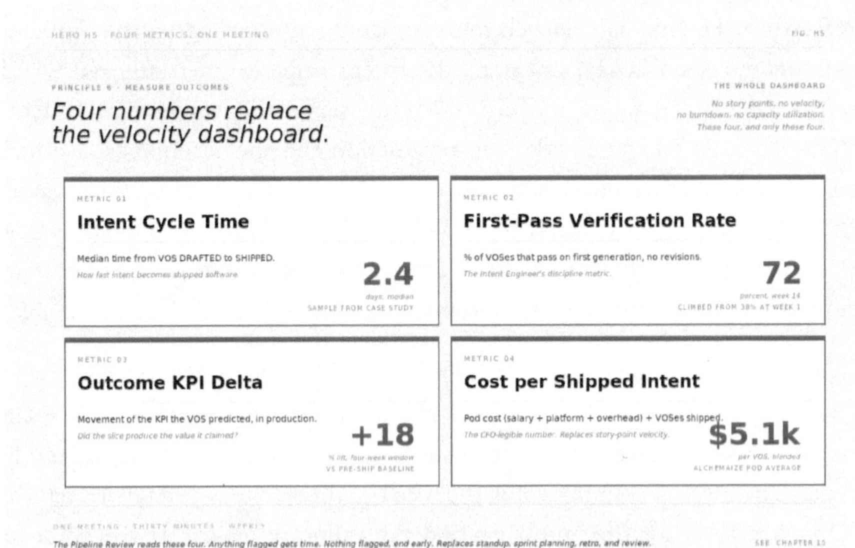

The full-page CID dashboard with all four numbers at every altitude.

Each one deserves a careful definition and an honest range. A metric without a range is decorative. A metric without a definition is political. The definitions below are the ones Alchemaize uses. The ranges are the ones we observed across the fourteen-week sprint from January through April 2026, and validated against the first three enterprise pilots that ran through April.

The shape of these four should look familiar to anyone who has read *Accelerate* [1]. Intent cycle time is the CID analogue of lead time for changes. First-pass verification rate plays the role that change failure rate played in the DORA work, inverted to a leading indicator and moved upstream of production. Outcome KPI delta is the variable the DORA research treated as the dependent one, the business outcome the delivery capabilities were meant to move. Cost per shipped intent is the efficiency term the DORA work did not emphasize, added here because generation economics make unit cost legible in a way it was not in a typing-bound pipeline. The four CID numbers map onto delivery capabilities, not ceremony outputs. That is the lineage, and it is deliberate.

Intent cycle time

Definition. The median elapsed time, measured in hours, from the moment a VOS enters the DRAFTED state (the Intent Engineer has authored it and committed it to the queue) to the moment the same VOS reaches SHIPPED-AND-OBSERVED (the code is in production and the acceptance contract has run against production behavior at least once). The start of the clock is not "when the idea occurred to somebody." The end of the clock is not "when the pull request merged." The clock runs from a specific state transition to another specific state transition, and both transitions are logged.

Why median, not mean. The distribution of cycle time in a healthy pod is heavily right-skewed. Most VOSes ship in a predictable window. A small number hit a verification failure or a context issue and take significantly longer. The mean is dominated by the tail. The median tells you what a typical VOS takes, which is the number a pod can calibrate against. If you want to track the tail, track the 90th percentile separately. Do not let the tail pollute the headline.

Honest range.

- Week 4: 36 to 48 hours. A pod that has run the pipeline for a month is still learning where its gates sit and how to size VOSes. A sub-24-hour cycle time in week 4 is almost always a sign that the VOSes are too thin, not that the pod is unusually fast. More on that in the diagnostic section and in Casey's field note.
- Week 12: 18 to 24 hours. Calibrated. The pod has stopped arguing about VOS boundaries, the verification harness has caught up, and the generation step has settled into its actual cost.
- Year 1: 8 to 12 hours. Mature. A VOS drafted on Monday morning is in production by end-of-day. The pod's constraints are now intent clarity and verification coverage, both of which have become the work.

What this replaces. Velocity, lead time as traditionally measured, and any metric that includes queue time the CID pipeline eliminates by design. Intent cycle time is the headline speed metric of a CID pod, and it is the only speed metric the pod needs.

First-pass verification rate

Definition. The percentage of VOSes that pass their acceptance contract on the first generation run, without a rework pass. A rework pass is any iteration where the generated code fails a contract assertion and has to be regenerated or modified. A single failed test, a single missing edge case, a single wrong function signature, counts as a failed first pass. The bar is intentionally sharp. A 100 percent first-pass rate would mean the pod is writing its acceptance contracts trivially, which is its own failure mode. A 0 percent first-pass rate would mean the intent specifications are so vague that generation is effectively guessing. The range in between is where a healthy pod sits.

Why this metric matters more than defect density. Defect density measures the bugs a human user finds after the code ships. It is a trailing indicator with a long lag and a low signal-to-noise ratio. First-pass verification rate measures the alignment between what the Intent Engineer specified and what the generation layer produced, the first time, against a contract the Verification Owner trusts. It is a leading indicator with same-day resolution and high signal. If the rate drops week over week, the pod knows something is wrong now, not three months from now.

Honest range.

- Week 4: 30 to 50 percent. The learning curve. Intent specifications are still carrying ambiguity. Acceptance contracts are still getting calibrated. Half of all VOSes take at least one rework pass, and that is fine.
- Week 12: 55 to 70 percent. Calibrated. The pod has learned which kinds of intent need more specification effort up front and which kinds don't. The Verification Owner has built a library of contract patterns that catch the common failure modes before they ship.
- Year 1: 75 to 90 percent. Mature. A first-pass rate above 90 percent is worth investigating. It usually means the contracts have become too permissive, or the VOSes have become too small to fail. Both are regressions, not improvements.

What this replaces. Defect density, code coverage, and the quality gates that in most SAFe implementations have degraded into approval checkboxes.

Outcome KPI delta

Definition. The measured movement in the specific business metric the stream is pointed at, relative to the stream's baseline at funding time. Every stream in ELCID has exactly one outcome KPI. Not three. Not a balanced scorecard. One number that the stream was funded to move, and that the CFO agrees is the number the stream is accountable for. Outcome KPI delta is the weekly movement of that one number.

Why there is no canonical range. The range is whatever the business decided at the stream charter. A stream funded to reduce monthly customer support tickets by 40 percent has a different range than a stream funded to increase monthly retention by 3 percentage points. The range belongs to the stream. What the CIO tracks is the slope: is the number moving in the intended direction, at the intended rate, relative to the calendar? A flat outcome KPI delta against a high shipping rate is the most useful diagnostic the four numbers produce. It says: the pod is building the wrong things.

What this replaces. Feature completion percentage, OKR progress as traditionally self-reported, and every variation of "business value delivered" that ends up being an estimate from the team doing the work. Outcome KPI delta is a real number taken from a real source system, verified by someone who does not report into the pod.

Cost per shipped intent

Definition. Fully loaded stream spend for the period, divided by the number of VOSes shipped in the period. Fully loaded means: human compensation allocated to the stream, AI platform consumption (tokens, compute, model inference), tooling subscriptions, and infrastructure cost for the stream's runtime environment. Not allocated corporate overhead. Not a percentage of the CFO's salary. The direct, attributable cost of running the stream for a week, divided by the stream's output.

Why this inverts as the stream matures. Early in a stream's life the VOSes are large, the team is learning, and the cost per shipped intent is high. As the pod calibrates, VOS size drops, first-pass rate rises, and the cost per shipped intent falls, sometimes by an order of magnitude. The inversion is the signal. A stream where cost per intent is rising at month six is a stream with a process regression, a context-bundle problem, or a VOS-sizing drift. All three are diagnosable from the number.

Honest range.

- Early (first four weeks): $2,000 to $5,000 per VOS. The team is learning. VOSes are large because the pod has not yet developed the reflex to split them smaller.
- Calibrated (weeks five to twelve): $500 to $1,500 per VOS. The pod has found its rhythm. Most VOSes fit inside a half-day of intent authoring and a single generation run.
- Mature (year one): $200 to $1,000 per VOS. The pod is shipping multiple VOSes per week per member, the generation layer has become reliable, and the overhead per VOS is dominated by verification time rather than intent authoring.

What this replaces. Cost per story point (meaningless, because story points are arbitrary and non-comparable across pods). Cost per developer (measures headcount, not output). "Engineering efficiency" metrics that nobody in the room can define.

Metric	Week 4	Week 12	Year 1
Intent cycle time (median)	36 to 48 hours	18 to 24 hours	8 to 12 hours
First-pass verification rate	30 to 50%	55 to 70%	75 to 90%
Outcome KPI delta	Set by stream charter; track slope.	Same	Same
Cost per shipped intent	$2,000 to $5,000	$500 to $1,500	$200 to $1,000

Table T6 · Metric ranges by maturity. Intent cycle time and first-pass verification rate carry hard ranges because the pod's calibration produces them. Outcome KPI delta has no canonical range because the range belongs to the stream charter. Cost per shipped intent inverts as the stream matures.

The full dashboard

Four numbers. Per pod, per week. Aggregated per stream, per month. Rolled up to portfolio level, quarterly. Same metric at every altitude. The CIO sees the portfolio number. The stream architect sees the stream number. The pod sees the pod number. When the portfolio number moves, it is because one or more stream numbers moved. When a stream number moves, it is because one or more pod numbers moved. The causal chain is always visible. There is no ambiguity about which pod moved the portfolio.

The dashboard fits on a single page. I mean that literally. The full-page hero diagram in this chapter is the dashboard. Four metrics, three maturity windows each, the week-over-week trend, the 4-week moving average, and a flag column for anything that moved more than 20 percent from the moving average. If a metric has not moved enough to flag, the CIO does not need to look at it. If a metric has flagged, the CIO has a single page of context and one question to ask the relevant stream lead. Thirty seconds. Four numbers. One portfolio. No dashboard theater.

3. Reading the four as a diagnostic

The chapter's most useful section is this one. The four numbers are not a scorecard. They are a diagnostic instrument, and they are most informative read in pairs.

A single metric, read alone, tells you a direction. Two metrics, read together, tell you a cause. The CIO who learns to read the four numbers in pairs can sit in a thirty-minute review with a stream lead and arrive at the operating question within the first three minutes. That is the posture the instrument is designed to produce.

Here are the four diagnostic pairs I have used most often, in the order they have been most useful.

High cycle time with high verification rate

The pod is being careful and the work is piling up. Intent specifications are precise, the generation layer is producing clean code on the first pass, and the acceptance contracts are landing. And the median VOS is taking 40 hours when it should be taking 20.

Somewhere in the system, there is a queue. The queue is usually in one of three places: the verification step (the Verification Owner is a bottleneck against rising throughput), the context-bundle curation step (the Intent Engineer is spending too long preparing the input), or the intent-authoring step itself (the pod has taken on a run of genuinely difficult VOSes and is paying the real cost of reasoning about them).

The diagnostic question is: *where is the work waiting?* Not *why is the pod slow.* The pod is not slow. The pod is queuing. Find the queue.

Low cycle time with low verification rate

The pod is shipping fast and producing rework. Cycle time looks beautiful on the dashboard. First-pass rate is under 50 percent and trending down. The pod feels productive. The pod is producing slop.

This pattern is the one that fooled me twice in the first quarter of the fourteen-week sprint. A dropping cycle time looks like a win. It isn't always. Sometimes it is a sign that the VOSes have thinned below the minimum useful size, and the pod is shipping small, trivially-verified changes that do not accumulate into outcome movement. Sometimes it is a sign that the acceptance contracts have degraded, and the pod is passing verification against a contract that does not reflect the business intent.

The diagnostic question is: *are the VOSes still doing useful work?* The answer comes from pairing cycle time and first-pass rate with outcome KPI delta. If outcome is flat, the VOSes are not doing useful work, no matter how fast they ship.

High shipping rate with flat outcome KPI delta

The pod is shipping and the business number isn't moving. This is the pattern that retires careers. Three months of high throughput, four-decimal precision on cycle time, 80 percent first-pass, and the product metric that the stream was funded to move has not moved.

The cause is almost never in the pod. The cause is in the intent portfolio. Somebody, months earlier, drafted a stream charter with a hypothesis about which features would move the KPI, and the hypothesis is wrong. The pod has been executing the hypothesis faithfully. The hypothesis is what needed revision.

The diagnostic question is: *what is the stream actually pointed at?* Not *what was it pointed at in the charter.* What is it pointed at in this week's intents. Read the last eight VOSes and ask whether, taken as a group, they could plausibly have moved the KPI. If the answer is no, the intent portfolio needs a review, not the pod.

Cost per intent rising while other metrics are stable

This one is subtle, because cost per intent is the metric least visible to the pod itself. Cycle time and first-pass rate are immediate, concrete, and motivational. Cost per intent accumulates over weeks and is usually surfaced at the monthly review.

When cost per intent rises while cycle time and first-pass rate hold, something has changed about what a VOS costs to ship. The most common cause I have seen is context-bundle regression: the Intent Engineer has started including more context per VOS than the work requires, usually as a response to an earlier first-pass failure that got over-corrected. Generation cost rises. Verification time rises. Cycle time holds because the pod compensated by shipping the same number of VOSes at higher per-unit cost. The dashboard does not show the compensation. The cost-per-intent number does.

The diagnostic question is: *what changed in how we are preparing work?* This one goes to the Intent Engineer, not the pod at large. Context curation is a specific skill with a specific cost curve, and the answer is almost always in there.

A fifth pair worth knowing

Two metrics that look fine in isolation can still flag together. The pair I watch most carefully is a simultaneous, modest, multi-week drift in cycle time and first-pass rate, both in the direction of degradation, with outcome and cost holding steady. A 10 percent rise in cycle time. A 5 percent drop in first-pass. Both within the month-over-month noise band. Neither one would flag a single-metric alert.

That pattern is how pods go stale. The process has absorbed a new piece of friction (a slower verification gate, a newer generation model with different reliability properties, a team member on leave, a context-bundle pattern that has stopped matching the workload). Nothing has broken. Everything is working. The pod is measurably, gradually, becoming a worse version of itself. If the CIO is not reading the pairs, this pattern will not surface until it costs a quarter of outcome KPI delta.

The diagnostic question here is less specific than the other four. It is: *what is different this month.* And if the stream lead cannot answer it, the CIO should spend a working session with the pod until the answer is identified.

> **Diagnostic reading, in one sentence.** Four numbers answer four different questions; four pairs of numbers answer the questions about why the organization is doing what it is doing.

That is the pull-quote. I do not often write pull-quotes for my own chapters. This one has earned its keep across every pod I have watched run the methodology, and it will do the work of teaching the CIO how to read the dashboard faster than any amount of additional prose.

4. The Friday review

The meeting is thirty minutes. Once a week. Friday afternoon, in every Alchemaize pod. Four numbers. No slides. No status storytelling. Three people in the room, or three people on the video call, depending on the week: the Intent Engineer, the AI Orchestrator, and the Verification Owner. That's it.

The agenda is five items. Five minutes on the four numbers, which most weeks is a single trend glance. Ten minutes on VOS flow, which is a table of what shipped, what is in progress, and what is entering the queue. Ten minutes on flagged issues, only if a number flagged. Five minutes on adjustments for the next week, if any. Thirty minutes, hard stop.

The rule that governs the meeting is the simplest and the most counter-intuitive for anyone coming from a Scrum-and-SAFe background. If no metric is flagged and no VOS is blocked, the meeting ends early. It is permissible for the Friday review to take eight minutes. In a mature pod, that happens roughly one week in four. The meeting does not exist to prove the pod is working. The meeting exists to catch the weeks when the pod is not working. The instinct to fill the time with narrative is the single most damaging habit a pod can import from its previous methodology, and the Verification Owner has explicit permission, in Alchemaize's operating rules, to end the meeting when the agenda is done.

At the portfolio level, one meeting a month replaces everything SAFe used to do in the PI Planning cycle. Scaled Agile publishes the full PI Planning and inspect-and-adapt cadence at `scaledagile.com` [2]; the canonical event is two days of planning every eight to twelve weeks, plus a program-level inspect-and-adapt at the end of each increment, plus the system demos and ART syncs that sit between them. Ninety minutes, once every thirty days, is what replaces all of that. The same four numbers, aggregated across streams. Each stream lead presents three things: shipped, in flight, friction. No slides. The decision agenda is at the end: what to start, what to continue, what to sunset. The decisions are made in the room. If the right people are not present, the decisions defer. The meeting is over in ninety minutes.

The total meeting load for a CID pod operating inside an ELCID portfolio is the thirty-minute weekly Pipeline Review, plus a ninety-minute contribution to the monthly portfolio review, plus a pair of fifteen-minute intake conversations at the start of each new VOS. That is the entire ceremonial footprint. Under two hours per week per pod. In exchange for the thirty-two hours per sprint that Chapter 2 counted for a Scrum team inside a SAFe release train.

The saved time is not the point, though the CFO will like it. The point is that the saved time has been replaced by a single instrument that actually answers the CEO's question. The ceremonies did not answer the question. The dashboard did not answer the question. Four numbers, read in pairs, at a cadence that matches the pipeline, answer the question.

5. What the four numbers do not measure

I want to close this chapter on what the instrument cannot do, because an instrument that is described only by what it does catches fewer of the people who would misuse it than an instrument that is also described by what it doesn't.

The four numbers do not measure engineer morale. A pod can have excellent cycle time and a first-pass verification rate in the high eighties and still be a pod where two of the three members are burning out. The metrics will not catch it. They are not pointed at it. The CIO who tries to read morale off the four numbers will read it wrong, and will discover the error at an exit interview.

The four numbers do not measure architectural debt. A stream can ship a quarter of VOSes at excellent cost per intent and be quietly accumulating a set of design decisions that will require a rewrite in year two. The verification contracts catch functional regressions. They do not catch architectural decay.

That is the Verification Owner's separate job, tracked on a separate review, and not visible on the four-number dashboard.

The four numbers do not measure strategic alignment. A stream can be pointed at an outcome KPI, and move it, and be pointed at the wrong KPI entirely. If the company has decided that monthly retention is the metric worth moving, and the retention increase the stream produced comes from a cohort the company is trying to de-emphasize, the outcome KPI delta will look great and the strategic result will be zero. That is a leadership conversation, not a measurement one.

The four numbers do not measure talent. The ratio of what a specific engineer is capable of to what the pod is producing is not legible on the dashboard. The stream might be running at high throughput because one member is carrying a disproportionate load. The dashboard will not say so. The Verification Owner and the Intent Engineer, in their one-on-ones, will know. The dashboard will not.

Here is the honest statement. The four numbers tell the CIO whether the engineering organization is producing outcomes against the intent portfolio it has been given. They do not tell the CIO whether it is the right organization, pointed at the right portfolio, executing on the right strategy, with the right people, for the right reasons. Those questions require leadership. The four numbers give leadership the clean instrument it needs to have those conversations on a factual basis, rather than on the political basis that the old dashboards enabled.

An instrument that answered all of those questions would not be an instrument. It would be a replacement for leadership. No dashboard has ever been that, and the book does not claim one exists.

What the four numbers do is remove the excuse. For twenty-five years, CIOs have been able to sit in front of a board and say that the engineering investment is producing results because the burndown is clean and the predictability is up and the velocity is trending to plan. The board did not have the vocabulary to say otherwise. The four numbers remove the vocabulary gap. Cycle time, first-pass rate, outcome delta, cost per intent. Is each one moving in the right direction? If yes, the investment is producing results. If no, which one, and why, and what is the next move?

That conversation does not require a forty-two-chart dashboard. It requires four numbers and a CIO who knows how to read them in pairs. The methodology gives both.

The old meeting was a narrative performance. The new meeting is a reading of the instrument. One of them produced activity reporting. The other produces decisions.

Field Note from the Test Pilot

> *Casey Robinson, Intent Engineer*
>
> **The week I was measuring my pace against the wrong timeline**
>
> This was around week five or six of the sprint. I was deep into the Campaign & Quest System on The Trade Codex, which was VOS #4 and one of the features I was most excited about. The idea was to wrap trading education in a narrative-driven RPG experience: campaigns with levels, side quests, XP, badges, and a world bible. The first campaign, Initiate's Ascension, had twelve levels, and each level needed its own narrative, objectives, quest definitions, and question bank content.
>
> I was shipping fast. Really fast. The campaign content work was highly repeatable once I had the pattern down. Write the level narrative, define the quest objectives, set up the question bank entries, wire the progression logic. Each level was its own VOS, and each one was coming through in about four to six hours from intent to shipped. I was averaging close to two levels a day some days. The cycle time on my VOSes had dropped significantly from where it had been a few weeks earlier, and I remember feeling like I had really found my rhythm.

On one of our regular calls (David, Glenn, and I were all on) we were looking at the numbers together the way we usually do. Not a formal review, just the three of us talking through where things stood. The cycle time looked great. I'd shipped eight or nine campaign-related VOSes that week. Everything felt like it was moving.

Then David pulled up the outcome we'd been tracking for the campaign system, which was campaign completion rate, the percentage of users who started Campaign 1 and actually finished it. The number was flat. Not down, not up. Just flat. Same as the week before.

My first instinct was that the number would catch up. I had a theory about it. The campaign needed all twelve levels to be live before users could complete it, and I was only on level seven or eight at that point. Users were starting the campaign, getting to wherever I'd built to, and then there was nothing left to do. The completion rate couldn't move until the whole thing was there.

David's perspective was interesting. He said something like, *the speed is real, but the outcome is on a different timeline than the cycle time. You're measuring your pace against a metric that can't respond yet.* Glenn added that the same thing happens in platform work, you can ship infrastructure VOSes all week and the user-facing metric won't budge because the infrastructure isn't the thing the user sees.

It wasn't a correction. It was more of a realization we all came to together. The cycle time dropping was genuinely good news. I was getting faster at the work, and the quality was holding up. But the outcome KPI was telling us that speed alone doesn't move the number. The campaign needed to be complete before the metric could respond. I was measuring my pace against a timeline that hadn't arrived yet.

What I took from that conversation was something I think about a lot now. When you're shipping fast and the outcome isn't moving, there are really two possibilities. One is that the work has thinned. You're shipping small things that don't add up to anything. That's the pattern David describes earlier in this chapter, and it's a real failure mode. The other possibility is that the work is real but the outcome is on a longer timeline than the cycle time. The campaign content was real work. Every level I shipped was a level users would eventually play through. But the metric couldn't see it yet because the feature wasn't whole.

The four numbers help you tell the difference between those two cases. If the cycle time drops and the VOSes are still substantial (real features, real acceptance contracts, real verification), then the outcome is probably on a longer timeline and you keep going. If the cycle time drops and the VOSes have gotten thin (cosmetic changes, small tweaks, things that don't connect to the outcome), then the work has drifted and you need to resize.

In my case, the work was real. I finished the remaining levels over the next week and a half. Once all twelve were live, the completion rate started moving. It went from flat to about 22 percent completion within two weeks of the full campaign being available. The metric had been waiting for the feature to be whole. Once it was, the numbers caught up to the pace.

I keep that experience in mind now whenever I see a flat outcome next to a fast cycle time. The instinct is to worry. The better instinct is to ask: is the work real, and is the outcome on a timeline the metric can see yet? Sometimes the answer is yes, keep going. Sometimes the answer is no, resize. The four numbers, read together, help you figure out which one you're in.

That's the meeting. Four numbers, read in pairs, once a week for thirty minutes, with the people who can actually do something about what the numbers say. If you're the one running it, the most useful thing I can tell you is to trust what it shows you, and to remember that a flat outcome doesn't always mean the work is off. Sometimes it means the work isn't finished yet.

Notes

1. Nicole Forsgren, Jez Humble, and Gene Kim. *Accelerate: The Science of Lean Software and DevOps.* IT Revolution, 2018.
2. Scaled Agile, Inc. *SAFe 6.0 for Lean Enterprises.* Scaled Agile, 2023. scaledagile.com/safe-big-picture (accessed 2026-04-22).

CHAPTER 11: COMPLIANCE AT GENERATION TIME

---

1. The theater

Compliance, as most enterprises practice it, is a quarterly performance. The auditor shows up on a scheduled cadence, samples a slice of the last ninety days of work, writes a finding or two, files a remediation ticket, and leaves. The team works the ticket into the backlog. Three weeks of code that shipped under a now-flagged pattern get rewritten. The auditor returns the next quarter, samples the next slice, and the cycle runs again.

This model made sense when code was slow and expensive to produce. A ninety-day review window was reasonable, because ninety days of work was a modest stack to audit. The cost of being wrong was bounded by the rate at which a team of humans could type. If the auditor found a HIPAA issue in March, the pattern had maybe affected two or three features. You rolled them back. You moved on.

At AI-generation speed, the window becomes catastrophic. A pod shipping under CID produces something like five to ten times the functional surface area per week that the same pod produced before. Ninety days is no longer a modest stack. It is a quarter's worth of compounded exposure, and the auditor is sampling a body of work that did not exist three months ago. If the auditor finds the same kind of finding, the blast radius is not two or three features. It is thirty or sixty. The remediation stops being a sprint. It becomes the quarter.

The standard response inside a regulated enterprise is to tighten the audit cadence. Move from quarterly to monthly. Move from monthly to weekly. Hire more auditors. Buy a continuous-controls-monitoring tool that claims to run the sample every night.

Frameworks like SAFe formalize this posture at the portfolio layer through Lean Portfolio Management and compliance epics, queuing regulatory work as dedicated items inside the program backlog. The pattern is still audit-time compliance. The work ships. The epic tracks the remediation. The governance board reviews.

None of it solves the actual problem.

The actual problem is that the audit is running on a cycle slower than the generation pipeline by an order of magnitude. Sampling a stream that is moving faster than you can inspect it does not make the stream any less wrong. It only produces a different report about the wrongness.

The theater has a tell. When a regulated team adopts an AI coding tool, the first two weeks look like a win. Throughput climbs. The backlog clears. The CFO asks whether the next hire is still necessary.

Then the compliance office catches up on its review and files the backlog of findings.

The findings are not small. The team spends the next quarter remediating the work that shipped in the first two weeks. Net output for the cycle is roughly zero. Sometimes it is negative, because the remediations themselves have defects.

I have watched this pattern end AI-coding pilots inside more than one regulated firm in the last year. The pilots do not fail because the AI was bad. They fail because the compliance model was still the old one.

The model was not designed for this equilibrium. It was designed for a typing-bound world. Pick up any compliance textbook written before 2024 and you will find the same assumption: verification runs downstream of a slow authoring step. Reverse the speed of the authoring step and the downstream verification is no longer a buffer. It is a leak. The theater keeps running because nobody has told the auditor their job is now structurally impossible. Most regulated firms respond by asking the auditor to work harder. That is the wrong request.

The right request is to move the verification.

> "A regulation is a predicate. If you can state it precisely enough to fine someone for violating it, you can state it precisely enough to block a commit that would violate it. Everything between those two statements is a choice about when the check runs."

2. The flip

Compliance at generation time means the rule is loaded into the workflow as a blocking constraint before the AI writes a line of code. Not a reviewer checklist. Not an audit finding. A predicate the machine evaluates at every gate it passes through, with the authority to stop the gate from opening.

The architecture is simple. A compliance extension is a set of numbered rules, each one mapped to a specific regulatory citation and paired with verification criteria stated precisely enough for the machine to check. The extension is installed into a stream at the enterprise governance layer. Once it is installed, the extension fires five times across the lifecycle of every VOS in that stream.

First, at VOS start, the extension presents an opt-in questionnaire. Does this feature handle Protected Health Information (PHI). Does it deploy into a FedRAMP-authorized environment. Does it process cardholder data. The Intent Engineer answers, and the answers determine which rule set activates for this VOS. Context stays lean for the VOSes that do not need compliance enforcement, which matters, because a pod does not want to pay the rule-evaluation tax on a feature that never touches a regulated datum.

Second, at requirements validation, the extension checks that the acceptance contract addresses the required compliance elements. For HIPAA: is there a PHI data inventory. Is there an encryption-at-rest specification. Is there an audit-trail scenario in the Gherkin. If any required element is missing, the VOS cannot transition out of CONTRACTED. The Verification Owner sees the gap. The Intent Engineer fills it. The contract advances.

Third, at generation time, every file the AI writes is checked against the active rules before the file is committed. For HIPAA-04, the audit-trail rule: is every PHI access in this file paired with an audit-log write. If the AI produces a data-fetch function that returns a patient record without a corresponding log entry, the rule fires. The commit is blocked. The finding is written to the audit log. The generator gets the finding back as input and regenerates.

Fourth, at verification time, the acceptance-contract scenarios that encode the compliance rule run alongside the functional scenarios in the harness. The audit-trail scenario either passes or it does not. If it does not, the VOS does not transition to SHIPPED.

Fifth, at observation time, once the VOS is in production, the rule continues to fire against runtime telemetry. If a PHI access in production lands without a corresponding audit entry, the observer raises a finding and the ticket is opened automatically. This is the only stage where the model tolerates a real-time finding instead of a blocking one, because by definition the code is already running. The job at this stage is to detect the drift and close it.

Five stages. One rule. The rule is the same object across all five. What changes is the moment and the medium. At VOS start the rule is a question to the author. At requirements validation it is a review criterion against the contract. At generation time it is a predicate against the file. At verification time it is a scenario in the harness. At observation time it is a check against

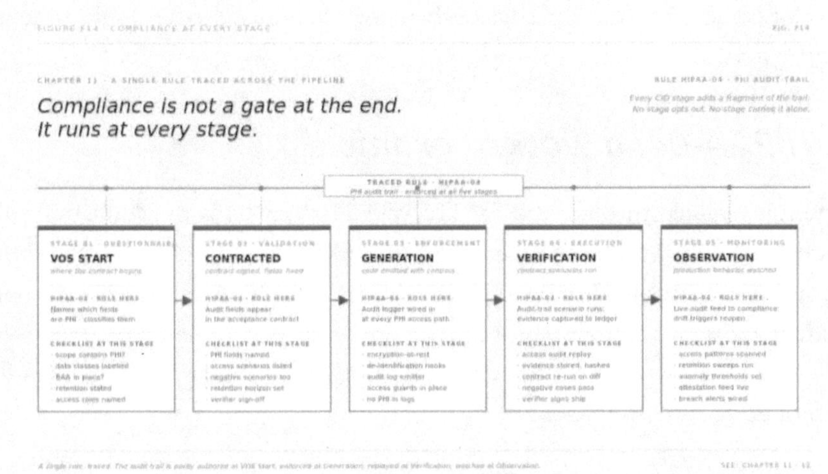

telemetry. A single compliance-enforcement object that activates five times, and fails closed at every activation.

Figure F14 · Compliance at every stage. *How a single rule fires across the CID pipeline at every stage: Intent, Context, Generation, Verification, plus the watching layer.*

The developer cannot override a blocking finding. The Verification Owner cannot either, under normal operation. The only path around a blocking finding is to demonstrate that the finding is a false positive, which updates the rule rather than skipping it. An override, if it ever exists, is logged, signed, and escalated outside the delivery chain. In practice, the override path almost never runs, because the acceptance contracts are tight enough that the legitimate cases are already covered, and the illegitimate cases should not ship.

The blocking message cites its source. When HIPAA-04 fires, the finding reads HIPAA-04, references 45 CFR §164.312(b), and describes the specific deficiency in the generated code. The developer who reads the finding gets both the rule and the citation. They do not have to look anything up. They do not have to guess whether the block is a stylistic preference or a federal statute. The citation is in the block.

This is the flip. The rule is not a document the auditors read and the developers ignore. It is a predicate the generator must satisfy, every time, at every gate.

3. HIPAA-04: a worked example

One rule, walked through the full lifecycle. HIPAA-04 is the audit-trail rule under the HIPAA Security Rule. Full citation: 45 CFR §164.312(b). The short form: every access to Protected Health Information must produce a tamper-evident audit entry, retained for at least six years, queryable by patient, user, and time, and written to append-only storage so an application identity cannot erase its own tracks.

Chapter 6 used this same rule as a demonstration of what an acceptance contract looks like. I am not going to repeat that contract here. What I want to show is what happens to the rule across the five enforcement moments, end to end, when a pod is building a feature that touches PHI.

Pick a concrete feature. The pod is adding a new endpoint to a clinical system that lets a care-team nurse look up a patient's most recent lab result. The endpoint is GET /patients/:id/labs/latest. It returns a lab record, which qualifies as PHI, to a clinician's session, which requires authorization. The VOS is authored by Casey, who has been on the Intent Engineer role for about six weeks.

Moment one, at VOS start. Casey opens a new VOS, and the compliance extension asks its opt-in questions. Does this feature handle PHI. Yes. Does it write to the append-only audit store. The question is pre-populated based on the stream's configuration, and Casey confirms. HIPAA activates. The VOS record is tagged with a compliance scope. The context pack the AI Orchestrator assembles now includes the HIPAA rule set, which adds about six kilobytes of regulation-mapped constraints to the model's working memory for this VOS.

Moment two, at requirements validation. Casey drafts the acceptance contract. His first pass has three scenarios: the happy path where the nurse fetches the lab, an authorization-denied scenario where the nurse is not on the care team, and an error scenario where the lab service is unavailable. The extension scans the contract for HIPAA-04 required elements. It finds the happy path. It finds the denial path. It does not find an audit-trail assertion in either. It raises a blocking finding at the requirements gate: *HIPAA-04: The acceptance contract does not assert that an audit entry is written for successful PHI access. Per 45 CFR §164.312(b), every access to PHI must be logged to an immutable audit trail.* The VOS does not advance out of CONTRACTED. Casey reads the finding, rewrites the scenarios to include the audit-entry assertions, adds the nine required fields in a table, and resubmits. The extension re-runs. The contract now satisfies HIPAA-04 at the contract layer. The gate opens.

Moment three, at generation time. The AI Orchestrator issues the generation task against the contract. The model writes the endpoint handler. On the first pass, it produces a clean-looking function that queries the lab service, filters the response by the caller's role, and returns the result. The handler is syntactically correct. It runs. It would pass a casual code review. It does not write to the audit store.

The extension inspects the generated file, applies the HIPAA-04 check, and finds a PHI access without a paired audit write. The commit is blocked. The finding is written to aidlc-docs/audit.md with the rule identifier, the regulation citation, the file path, and the offending line numbers. The Orchestrator receives the finding as the next input. The next generation produces a handler that calls the audit-log service before it returns the lab record, with the field set the rule requires. The extension re-runs. No findings. The commit proceeds.

Elapsed wall-clock time from the first blocked commit to the successful commit: under two minutes. Elapsed developer-hours: zero, because no human wrote the fix. The Orchestrator read the finding, routed it back into the generator, and the machine produced the correct version on the second attempt. Casey reviewed the result and moved on.

Moment four, at verification time. The harness runs the acceptance contract. Every scenario executes as an automated test. The happy-path scenario runs, and its assertions check that the lab record is returned and that an audit entry was written within 250 milliseconds with all nine fields populated. The denied scenario runs, and its assertions check that the 403 was returned and that an audit entry was written with the denial outcome. The audit-write-failure scenario runs, and its assertions check that the endpoint returns 503 and does not disclose the lab record when the audit store is unavailable. All three pass. The VOS is eligible for the Verification Owner's signature.

Moment five, at observation time. The VOS ships. The endpoint goes live in a care team's production environment on Tuesday afternoon. On Thursday morning, the observer, which is a background service the Verification Owner maintains, runs a reconciliation query against the audit store. For every PHI access logged by the application, there should be a corresponding entry in the append-only store. A discrepancy would fire an alert. None does. The VOS is clean. Six weeks later the rule catches something else, which I will come back to in Section 5.

That is the full sweep. One rule. Five activations. Every activation fails closed. No human read the statute, and no human had to. The citation was loaded into the extension once, by whoever authored the extension, and it did its job every time the feature touched PHI from that point forward.

The part that tends to surprise regulated-industry CTOs the first time they see this is that the compliance overhead, measured in developer minutes per VOS, is lower under this model than under any audit-based model they have run. The rule fires every time. The rule almost always passes, because the contract already encoded it. The one time the rule blocks, the block resolves in under two minutes with no human intervention.

The old model paid a large overhead at quarterly boundaries and carried a large expected cost for the findings the auditor would eventually uncover. The new model pays a tiny overhead continuously and carries near-zero expected cost for post-hoc findings, because there are almost no post-hoc findings to uncover.

Stage	The rule's medium	What the rule does	Failure posture
1. VOS start	Opt-in question to the author	Determines which rule sets activate for this VOS	Skipped if not applicable; cannot be silently bypassed
2. Requirements validation	Review criterion against the acceptance contract	Confirms required compliance elements are present in the Gherkin	VOS cannot leave CONTRACTED until gaps are filled
3. Generation time	Predicate against each generated file	Checks the file against active rules before commit	Commit blocked; finding fed back for regeneration
4. Verification time	Scenarios in the acceptance harness	Compliance scenarios run alongside functional scenarios	VOS cannot transition to SHIPPED if compliance scenarios fail
5. Observation time	Check against runtime telemetry	Watches production for drift from the rule	Real-time finding raised; ticket opened

Table T7 · Compliance by stage. A single rule across five activations. The medium changes; the rule does not. Stages one through four fail closed. Stage five detects drift and routes it.

4. The three extensions

Three compliance extensions ship with the current release. Twenty-two rules in total. Each rule is numbered, each rule cites its source regulation, and each rule fires at one or more of the five activation stages. The rule documents are public, in the `cid` repository under `extensions/compliance/`. The runner that loads them and enforces them at generation time ships with CATALYST.

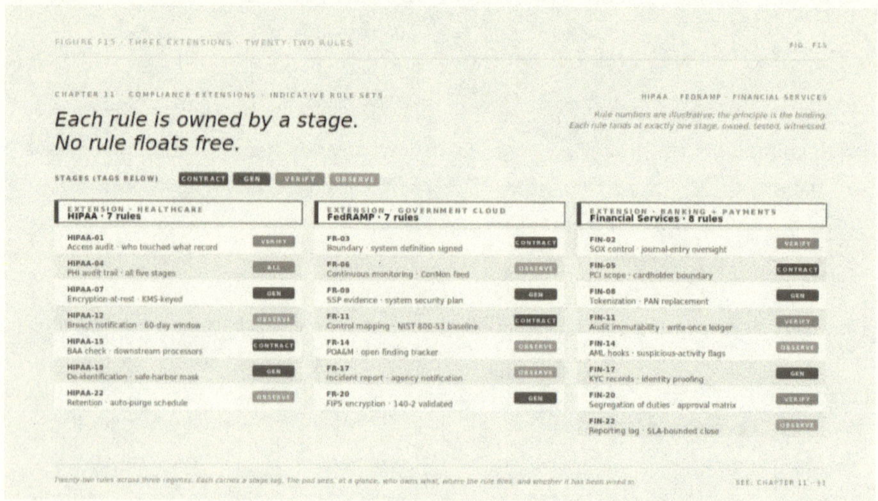

Figure F15 · 3 extensions, 22 rules. *HIPAA, FedRAMP, and Financial Services extensions with their rule counts.*

HIPAA. Seven rules, HIPAA-01 through HIPAA-07. They cover PHI classification and inventory, minimum-necessary access, audit trail, encryption at rest and in transit, Business Associate Agreement verification for external services, breach-notification readiness, and access logging for third-party integrations. Each rule maps to a specific section of 45 CFR Part 164 Subpart C. HIPAA-04, the audit-trail rule used in Section 3, maps to §164.312(b). HIPAA-01, the inventory rule, maps to §164.514(a). The mapping is in the rule file. An auditor can trace from the rule identifier to the citation in one step.

FedRAMP. Eight rules, FEDRAMP-01 through FEDRAMP-08. They cover boundary definition, FIPS 140-2 cryptography, continuous monitoring, access control baselines, configuration management, incident response readiness, supply-chain integrity, and authorization artifact maintenance. Each rule maps to a control family in NIST SP 800-53 Rev 5. The FedRAMP extension carries a scope tag on each rule that determines which impact level it applies to: [All] for Low, Moderate, and High; [Moderate+] for Moderate and High; [High] for High only. The opt-in questionnaire asks which impact level the system is targeting, and the rule set activates accordingly.

Financial Services. Seven rules, FINSVC-01 through FINSVC-07. They cover sensitive-financial-data classification, cardholder-data protection under PCI-DSS v4.0, financial-reporting integrity under SOX §404, SOC 2 trust-services criteria, change-management audit trails, key management, and retention periods. Each rule carries a scope tag: [PCI], [SOX], or [ALL]. The opt-in questionnaire asks which regulations are in scope for the stream, and the applicable rules activate. A stream that handles cardholder data but no SOX-reportable financials activates the PCI rules and skips the SOX rules. A stream that reports publicly-traded financial results activates SOX and skips PCI. A stream that does both activates both.

Twenty-two rules is not a large rule set. It is a deliberately small one. Each rule is written to be independently verifiable, which means the verification criteria have to be concrete, measurable, and separable from every other rule. A rule that depends on another rule for its own evaluation is not really a rule. It is a sequence, and sequences break when the order of operations changes. Keeping the rules independent is one of the main editorial jobs of the extension author.

The rules are not designed to cover every regulatory obligation under every regime. HIPAA has administrative safeguards that do not reduce to a code-level predicate, and the extension does not try to enforce them, because they do not live in the software. SOX has financial-control obligations that belong to the finance organization, not the engineering organization, and the extension does not try to enforce those either. The principle is straightforward. If the regulation reduces to something a generator or a harness can check, it goes in. If it does not, it belongs in the human compliance program, and the extension does not pretend to cover it. The line is bright, and it is the line David and I have spent the most time arguing about when we add a new rule.

The three extensions cover a meaningful share of the regulated-industry stack by market. Healthcare runs on HIPAA. Federal workloads run on FedRAMP. Banking, brokerage, card-present retail, and publicly-traded companies run on some combination of PCI-DSS, SOX, GLBA, and SOC 2, all of which are covered by the Financial Services extension. A banking-specific extension sits in the repo but is not part of the standard release, because the sector-specific overlay on top of the Financial Services extension is thin enough that most banks run the Financial Services extension with a small custom overlay rather than a separate extension. That is a pattern I will return to in Section 6.

One note on the rule-numbering convention. Rule IDs are stable across releases. If HIPAA-04 means audit-trail in the April 2026 release, it still means audit-trail in the October 2026 release, even if the verification criteria have been tightened or expanded. An auditor who references HIPAA-04 in a finding knows they are referencing the audit-trail rule, regardless of which version of the extension was active at the time. Stability of identifiers is a boring discipline that matters every time a third party must cite the system years after the fact.

5. The ROI argument

The dollar argument writes itself. A single HIPAA violation carries a civil monetary penalty that starts at $50,000 per violation, tops out at $1.5 million per identical provision per year, and can be layered across multiple provisions for a single breach. A single PCI-DSS breach carries reputational cost, card-brand fines, and bank fees that routinely run into seven figures, and in a few public cases eight or nine. A single SOX material weakness disclosed in a 10-

K can move a public company's stock by double-digit percentages in a week. One penalty under any of the three regimes funds an entire CATALYST engagement with margin to spare. This is the argument the CFO hears first, and it is the argument the board is easiest to defend to.

It is also not the argument that matters.

The one that matters is the capacity argument.

Regulated engineering teams routinely lose 20 to 30 percent of their shipping capacity to post-hoc compliance remediation. This is not a number I made up. It is a range I have measured across six regulated-industry consulting engagements in the last fourteen months, and it matches internal audit-burden estimates published by several of the top-ten health systems and at least two of the major banks. The work is real. Compliance files a finding. The team triages the finding. The team writes a remediation ticket. The ticket competes with new-feature backlog for sprint capacity. The team ships fewer features, or the team ships the same number of features by skipping something else, which produces the next round of findings, which produces the next round of remediation. It is a treadmill.

Compliance at generation time makes the treadmill disappear. Not because the team is suddenly shipping non-compliant code and getting away with it. Because the code is compliant when it ships, and there is nothing to remediate in the next quarter. The capacity that used to pay for post-hoc remediation is now available for forward work.

Run the number at a specific team size. A twenty-person regulated engineering organization at the median fully-loaded cost of $250,000 per year per engineer is a $5 million annual capacity. Twenty-five percent of that is $1.25 million in recovered capacity per year. That is the number the CFO should write on the whiteboard when the investment is being evaluated. A CATALYST engagement with compliance extensions, amortized across the first year of adoption, costs well under that $1.25 million at any reasonable team size. The engagement pays for itself on capacity recovery alone, before the first penalty is avoided, and before the first cycle-time improvement is booked. The compliance argument is net-positive inside the first year without needing the risk-avoidance case at all.

The risk-avoidance case is still worth making, because the tail is fat. A single nine-figure breach in a regulated firm is a career-ending event for the CIO and the CISO, and it can be existential for a mid-cap company. The expected value of avoiding the tail is asymmetric. The cost of the compliance investment is bounded. The cost of a single failure mode at the tail is essentially unbounded. A CFO who discounts the tail because it has not happened at their firm yet is doing a present-value calculation that ignores the distribution. Compliance at generation time is a hedge against the tail, and it is paid for by the median-case capacity recovery.

There is one more number worth naming. The Verification Owner's signature at SHIPPED is a trust anchor that extends to the auditor. An auditor who can cryptographically verify that every shipped VOS passed a specific rule, in a specific release of a specific extension, against a specific regulatory citation, can shorten their own work dramatically. Sampling is no longer necessary when the entire population has been verified. Several CATALYST customers have already negotiated shorter audit engagements on this basis. Two have renegotiated their continuous-monitoring service costs downward. The savings on the audit side is harder to quantify in a general way, because every firm has a different audit footprint, but it is real, and it compounds.

The quarterly audit theater costs money to run. It costs more money when it finds something. It costs even more money when it misses something and the regulator finds it instead. Compliance at generation time collapses all three cost centers into a single, small, continuous overhead that falls inside the engineering pod. That is the argument the book is making, and it is the argument a skeptical CFO can defend in the boardroom with a straight face.

6. Custom extensions

Most regulated organizations eventually need a compliance extension that is not in the standard release. A bank with its own internal model-risk-management policy. A hospital system with a state-level privacy overlay that goes beyond HIPAA. A defense contractor with an ITAR or CMMC requirement that sits on top of FedRAMP. A publicly-traded company with a corporate policy on third-party data sharing that is stricter than any applicable regulation. The standard release cannot cover all of these, and it does not try to.

Authoring a custom extension is an authoring exercise, not an engineering project. That distinction matters, because it determines who does the work and how long it takes.

An engineering project is what the old vendor-compliance model used to deliver. A consulting firm would arrive, write a specification, build a custom rule engine, integrate it into the client's pipeline, and charge a six-figure fee. The delivery took three to six months. The rules were locked into the engine's proprietary format. Changes required a change order. Maintenance was a recurring annual contract. The whole shape of the engagement was an engineering shape.

An authoring exercise is different. The person who authors a CATALYST extension is someone who understands the regulation or the policy, can state each rule in one or two paragraphs, can specify the verification criteria precisely enough for a machine to check, and can cite the source authority for each rule. That person is typically a senior compliance officer, a risk lead, or a regulated-industry SME with a working grasp of how acceptance contracts are written. They do not need to be an engineer. They need to be able to write a specification that a generator can evaluate, which is a skill most senior compliance officers already have, because they have been writing specifications of that kind for regulators for years.

The time scale is a fortnight, not a quarter. An organization with a working CID pod and a senior compliance officer can draft a custom extension in two weeks. The structure is simple. A rules file with numbered rules, each one carrying a verification block and a regulatory citation. An opt-in file with the questions that activate the rule set. Both files are markdown. Both files live in the project alongside the code. No server. No SaaS. No vendor-locked format. If the organization's compliance program changes, the extension file changes, and the next VOS picks up the new rule on the next generation.

The authoring process mirrors the VOS process itself, which is part of why it works. The compliance rule is a specification. The generation-time enforcement is the executable verification. The author is doing, at the extension layer, the same move the Intent Engineer does at the feature layer: stating intent precisely enough for the machine to satisfy it. A compliance officer who has watched a pod run CID for a few weeks has already internalized the shape of the work. Writing an extension is the same discipline applied to a different artifact.

One thing I tell every organization considering a custom extension. Do not try to encode the whole regulation. Encode the parts that reduce to a predicate. The parts that do not reduce to a predicate belong in the human compliance program, and pretending they live in the extension is worse than leaving them out, because it gives the organization a false sense that the extension is enforcing something it is not. The extension is an engineering artifact. It enforces the parts of the regulation that are engineering concerns. The rest of the regulation is still the compliance office's job, and it is still the general counsel's job. The book is not trying to replace either of them.

A second thing I tell every organization. The extension is a living document. A regulation is a moving target. Enforcement guidance shifts. Precedent accumulates. The rules in the extension must track those changes, which means the Verification Owner, working with the compliance officer, must review the extension on a recurring schedule and update it when the underlying rule shifts. We run a quarterly extension review at every CATALYST customer that has regulated workloads. It takes about a half-day per extension. It is one of the most consequential half-days on the calendar.

The economics of custom extensions are favorable enough that the pattern tends to multiply. An organization that authors one extension almost always authors a second, because the team has figured out the authoring shape and the next regulation is easier than the first. By the end of a first year, a regulated CATALYST customer typically has two to four custom extensions in addition to the standard three, covering the specific overlays and internal policies their business requires. The pattern stabilizes there. The total rule count across all extensions at a mature customer is usually thirty to forty rules. Most VOSes activate no extensions. A few activate one. The rare VOS activates two or three. The cost of carrying the full set is borne once, in the authoring, and the marginal cost at generation time is close to zero.

7. The counterintuitive conclusion

Regulated industries are the best candidates for AI-native software development, not the worst.

The common assumption is the reverse. Regulated industries get named last on every AI-adoption list. The compliance burden is cited as the reason. The implicit argument is that the regulation makes the AI too risky to use. The argument is backwards. The regulation makes the AI the only sane option, because the regulation is precisely what compliance at generation time encodes. A regulated industry has already done most of the work of stating, in writing, what correct behavior looks like. The regulation is the specification. A CATALYST extension is the machine-evaluable form of that specification. The industries that have spent the last thirty years documenting their constraints in regulatory filings are the industries whose constraints are easiest to load into a generator.

This is not a theoretical claim. In the first fourteen weeks of Alchemaize's CID operation, four of the thirty-five applications shipped under the methodology touched regulated data. Two under HIPAA, one under PCI, one under SOX-adjacent reporting. The regulated ones shipped faster than the unregulated ones, because the acceptance contracts wrote themselves off the regulatory text, and the verification harness caught the failure modes a human reviewer would have spent weeks finding. The team's first-pass verification rate on regulated VOSes was higher, not lower, than on unregulated ones. The compliance office got more, not less, of what it was asking for, and got it sooner.

The inversion has consequences for the book's remaining chapters. Chapter 12 covers anti-patterns, and one of the most common is an organization that believes its regulatory footprint disqualifies it from AI-native adoption. It does not. The footprint is the advantage. Chapter 13 covers adoption, and the adoption path for a regulated firm is shorter than the adoption path for an unregulated one, because the regulated firm already has the compliance officer who will become the extension author on day one.

The chapters that follow assume the reader now holds this inversion. The compliance argument is not an afterthought to CID. It is the case that converts the skeptical enterprise from pilot to production. Teams that have taken the compliance argument seriously ship regulated software at AI speed and sleep through the quarterly audit. Teams that have not will stay on the old treadmill, running faster each quarter against a backlog of findings they cannot catch up to. The choice is in front of every regulated CTO reading this book. The model that works is not a harder version of the old one. It is the one that moves the rule to the moment of generation and lets the machine carry the weight.

Field Note from the Test Pilot

> *Casey Robinson, Intent Engineer*
>
> **The day I asked what happens when a user talks to our AI**
>
> This was around week four or five of the sprint. I was working on the AI Trade Analysis feature for The Trade Codex, which was VOS #3. The feature lets users complete a trade and get intelligent feedback from Bedrock, pattern recognition, risk assessment, suggestions for improvement. The user logs their trade, and the AI analyzes it and gives them a write-up.
>
> I was testing the feature after it shipped, clicking through it the way a user would, and I had one of those moments where you

see something from a different angle. The trade notes field. Users can type whatever they want into it. And whatever they type goes into the prompt that gets sent to Bedrock for analysis. I sat there looking at the screen and thought: what happens if someone types something in there that isn't about their trade?

I didn't know the term for what I was worried about. I learned later it's called prompt injection, where a user types instructions into a field that gets fed to an AI model, trying to get the model to ignore its original instructions and do something else. At the time, I just had a feeling. If users can type into a field that talks to AI, someone is going to try something creative. Maybe they try to get the AI to give them stock tips instead of trade analysis. Maybe they try to get it to reveal its system prompt. Maybe they just type something weird to see what happens.

I brought it up on our next call with David and Glenn. Not as a formal finding or a compliance issue. Just as a question: "What happens if a user types something into the trade notes that isn't about their trade, and it goes to the AI?" David's perspective was that this was a real concern, especially for a production app where we're responsible for what the AI says back to users. Glenn's perspective was more architectural. He pointed out that the AI's response goes back to the user's screen, so if someone could manipulate the prompt, they could potentially get the AI to say things we wouldn't want associated with our product.

We spent about an hour on it together. Not a formal security review. Just the three of us talking through the scenarios and figuring out what to do. We added guardrails to the AI prompts, system-level instructions that tell the model to stay in its lane and only discuss the trade data it's been given. We added input validation on the trade notes field to flag anything that looked like it was trying to redirect the AI. And we added a check on the AI's response to make sure it was actually about the trade and not about something the user tried to steer it toward.

Here is what I want to share about that experience.

I am not a security specialist. I did not know what prompt injection was before that conversation. I still couldn't write a technical specification for how to prevent it. What I could do was look at the feature from the user's perspective and ask a question that turned out to be the right question. The methodology gave us a way to turn that question into action. We updated the VOS, added the security scenarios to the acceptance contract, and the AI tool helped us implement the guardrails. The whole thing took about ninety minutes from the question to the shipped fix.

The part that stays with me is that the question came from thinking about the feature the way a user would think about it, not the way an engineer would think about it. An engineer might have caught the prompt injection risk from a technical angle. I caught it from a "what would a person actually do with this" angle. Both angles lead to the same fix. The methodology doesn't care which angle the question comes from. It cares that the question gets asked and that the acceptance contract captures the answer.

Glenn has talked earlier in this chapter about compliance extensions and how they catch things automatically at generation time. I think that's powerful, and for regulated industries it's essential. What I want to add from my perspective is that not every security concern comes from a regulation. Some of them come from someone on the pod looking at the feature and asking a simple question. The methodology supports both. The formal compliance extensions catch the things you're required to catch. The pod's own judgment catches the things nobody wrote a regulation for yet. Both matter, and the methodology gives you a place to put both of them, in the acceptance contract, where they become part of what gets verified before anything goes to users.

I wrote a lot of VOSes in the hundred days. A handful of them had security considerations I wouldn't have thought of on my own. Every time one of us raised a concern, we talked it through as a pod, updated the contract, and moved on. The methodology made that natural.

That's the thing I'd want a reader to take from this. You don't have to be a security expert to catch a security issue. You have to be willing to ask the question, and you have to be working inside a methodology that gives you a place to put the answer.

PART IV

The Road

Chapters 12 through 14.

What gets in the way, how to start, and who builds next. Part IV is the practical half of the book. The anti-patterns name the failures the methodology produces when it fails. The adoption path names the smallest first step. The closing chapter names the readers the methodology was actually built for.

CHAPTER 12: THE COSTUMES

Glenn Knepp

---

1. What you wear and what you do

An org chart is a hat rack. Every head gets a hat. Every hat has a title stitched into it. VP of Engineering. Engineering Manager. Scrum Master. QA Lead. Tech Lead. Business Analyst. Project Manager. Release Manager. Director of Delivery. The hats go on in the morning, come off at night, get passed around at reorgs, and get renamed every time a consulting firm comes through the building with a deck.

The hats are real. The work those hats describe is not always real anymore.

Walk onto a CID pod at month three. Three people on the call. One is writing intent. One is running the model. One is signing the verification. Ask the three of them what their job titles are, and you will hear six or seven titles between them. An Intent Engineer who used to be a Scrum Master. An AI

Orchestrator who used to be a Tech Lead, and before that a Senior Developer. A Verification Owner who used to be a QA Lead, and before that a test architect in a defense contractor. The titles on the HR record have nothing to do with the seat they are sitting in. They never quite did. They are doing so even less now.

That is the argument of this chapter. Org-chart titles in an organization running CID become costumes. People wear them because the HR system requires a hat. The actual work is allocated by pod role, and the pod role is almost never the same thing as the title. Costumes are not useless. A costume gets you paid on a band. It gets you a line on the org chart. It gets you into the meeting when the CFO asks who is running engineering. It does not, by itself, tell you what anyone is doing.

I want to be careful here. I am not saying the people holding these titles are useless. I am not saying the function the title used to cover is useless. I am saying the function is now done somewhere else, or done differently, or done by a pod that does not need the title to do it. The hat is still on the head. The work has moved.

This chapter names the titles that have become costumes, says where the work went, and says what to do about the person still wearing the hat. It is diagnostic, and a little flat. The subject earns it.

2. Three seats, everything else optional

CID has three pod roles. Intent Engineer. AI Orchestrator. Verification Owner. Three per pod. That is the whole role chart.

The Intent Engineer writes the intent. That means the WHY, the WHAT in Gherkin, the scenarios, and the outcome hypothesis. The Intent Engineer is closer to a product person than an engineer, and closer to a technical writer than either, and is usually neither of those things on their prior resume.

The AI Orchestrator operates the model. That means context curation, task-plan review, sub-task approval, mid-generation intervention, and the judgment call about when to regenerate and when to edit. The AI Orchestrator is the most senior technical seat on the pod, and the one most often misfilled when a company tries to run CID with the first engineer who volunteered.

The Verification Owner owns the acceptance-contract standard, the harness, the rollback path, and the signature. The seat reports outside delivery. Chapter 6 covered the role at length, and I will not retread it here.

Three seats. Three functions. No standing architect. No standing tech lead. No standing QA. No standing scrum facilitator. No standing project manager. If a pod has a fourth person, that person is a specialist brought in for a bounded piece of work, usually compliance or security or a domain expert for a specific VOS, and they leave when the VOS ships. The pod does not grow a permanent fourth seat. It never needs to.

The reason it never needs to is that the old seats existed to handle coordination problems that are now handled in the artifact. A backlog that is always visible does not need a person whose job is to tell the team what is in the backlog. A verification signature that happens on every ship does not need a person whose job is to convene the meeting where shipping is blessed. A specification that is authored in Gherkin before the code is written does not need a translator between the business and the engineers. The artifacts carry the coordination. The coordination roles become costumes.

What follows is a tour of the common costumes, what the title used to do, and where the work went.

3. The costume gallery

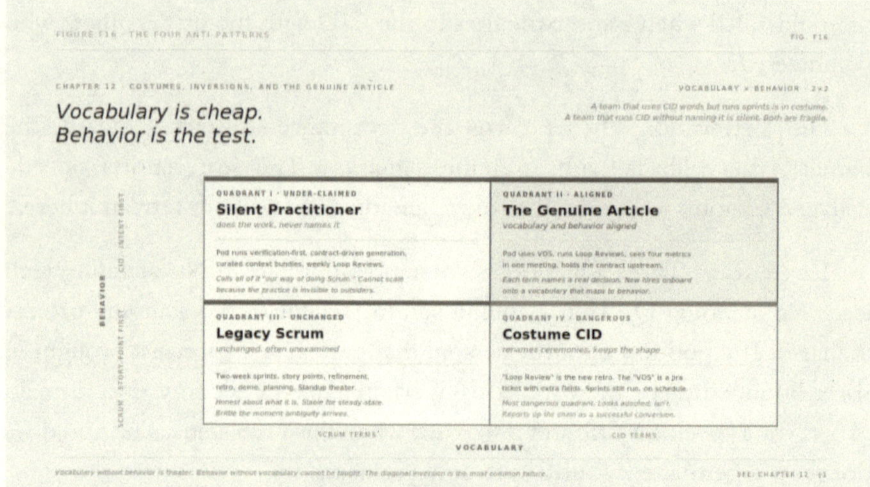

Figure F16 · Anti-pattern quadrant. *The common costumes mapped against where the work went, and how the role converts.*

VP of Engineering

The title was invented to solve a real problem. At some company size, engineers needed a person whose full-time job was to hire, fire, promote, budget, and protect the engineering function from the rest of the business. The VP of Engineering sat between the CEO and the engineering line, translating business pressure into technical priority, and technical reality into business language. In a hundred-person engineering organization running Scrum, this is a genuine job.

In an organization running CID, the translation is done by the artifact. The intent stream has a named owner, a budgeted run rate, a verifiable outcome, and a monthly review. A CFO who wants to know what engineering is doing reads the stream dashboard. She does not need a VP to narrate it for her. The hiring problem is still real. The budget problem is still real. The protect-the-function problem is still real. Those functions persist. They just do not require a full-time seat at the VP band to carry them out.

The hiring is usually done by the Verification Owner, who knows what good looks like, in partnership with the Intent Portfolio Lead at the enterprise layer. The budget is done by finance against the stream run rate. The protect-the-function problem mostly evaporates, because the function is no longer a queue of typing that must be defended against interruption.

The person wearing the VP of Engineering hat is still on payroll. Most of them are now doing one of three things. Running a stream as Intent Portfolio Lead. Sitting in the Verification Officer seat at the enterprise layer. Or doing the actual hiring and people work that used to be crammed into the leftovers of their week. The third one is the most common. It turns out there was always a full-time people job inside the VP of Engineering role, and removing the coordination load revealed it. The person is busier doing fewer things.

Engineering Manager

Engineering Manager was the seat that held a team of six to ten engineers together through the weekly rhythm of a sprint. Ran the standups. Sat in the sprint planning. Owned the retrospective. Carried the one-on-ones. Was the escalation for the Scrum Master when the Scrum Master needed muscle. In most organizations, the Engineering Manager was a former engineer who had been promoted into a coordinator role and spent eighty percent of their week in meetings.

The sprint is gone. The standup is gone. The sprint planning and the retro are gone. The Pipeline Review is thirty minutes a week. The coordination load on the Engineering Manager has dropped by roughly an order of magnitude, and the work that is left is the one-on-ones and the career development, which is roughly a quarter-time job for the size of team the Engineering Manager used to run.

The person wearing the Engineering Manager hat usually ends up in one of two places. Either they take the AI Orchestrator seat on a pod and remember they were an engineer, which they mostly like. Or they pick up responsibility across two or three pods as a people manager and a mentor, which does not need a seat on the pod and does not need a daily meeting.

The "manager of two pods with no meetings" seat is the hardest one for a traditional Engineering Manager to adjust to. The job used to be structured by the meeting calendar. Now there is no meeting calendar, and the work has to be structured by outcome rather than by cadence. Some of them take to it. Some of them do not, and take the AI Orchestrator seat instead, which is a real engineering role and pays the same.

Scrum Master

Scrum Master is the purest costume case in the gallery, because the ceremonies the role was invented to run are the ceremonies CID most deliberately eliminates. No standup to facilitate. No sprint planning to run. No retro to hold the space for. No velocity chart to update. The role's calendar empties.

It is worth being precise about what is being eliminated, because the costumes in most shops bear only a family resemblance to the prescriptions they cite.

The Scrum Guide describes the Daily Scrum as a 15-minute event for developers to inspect progress toward the sprint goal and adapt the plan for the next day's work. Nowhere does it prescribe a status round delivered to a manager. SAFe describes PI Planning as a cadence-based event for the Agile Release Train to plan work for the next program increment, typically eight to twelve weeks.

Whatever the ceremonies have become in practice, the costume the Scrum Master is asked to run rarely matches either prescription.

What CID retires is the costume. The prescription was already gone before CID arrived.

I watched this in three engagements last year, and every time the conversation with the Scrum Masters went the same way. The first response is denial. The ceremonies must still be running somewhere, and they must still be needed. The second response is grief. The calendar is empty, and the work the person was trained for is gone. The third response, if the organization handles the conversion well, is that the Scrum Master starts looking at Intent Engineer job specs and realizes most of the craft ports over. The Scrum Master was already reading the backlog, already interrogating vague stories, already pushing back on product owners who said "make it better" without saying what better meant. All of that is Intent Engineer work. Swap Jira for a VOS template and Gherkin for a user story, and the skill curve is measured in weeks, not quarters.

The organizations that do not handle the conversion well keep the Scrum Masters, give them a rebranded title, and ask them to "facilitate CID ceremonies" that do not exist. The Scrum Master, understandably, starts inventing ceremonies to facilitate. That is the Ceremony Creep anti-pattern from Chapter 12's diagnostic boxes, and the root cause is almost always a Scrum Master kept in place with nothing to do. Give them the Intent Engineer path or part ways. Do not leave them in the middle.

QA Lead

QA Lead ran the test lane at the end of the pipeline. Owned the test plan, the test environments, the test data, and the release sign-off. In a regulated environment, the QA Lead was often the person whose signature legally mattered, because the regulator wanted a human attestation before the artifact went to production.

The function has moved upstream. The acceptance contract is the test plan. The Gherkin is the test data specification. The harness runs the tests. The Verification Owner signs the release. All four used to be the QA Lead's work. The first three are now artifacts, and the fourth is the Verification Owner role.

In almost every CID adoption I have consulted on, the best QA Lead in the organization becomes the first Verification Owner. They already know how to read a specification for holes. They already know what it feels like to be the last line of defense. They already know how to hold the standard when the delivery lead is working the calendar against them.

The promotion is real. Pay band moves. Reporting line moves out of delivery. Seniority moves up.

If the organization tries to fill the Verification Owner seat with the QA Lead at the QA Lead's old band and reporting line, the organization has renamed a role rather than adding one. That is the most common corruption of the Verification Owner seat I see in the field, and it is how a CID adoption silently fails. If the QA Lead is going to be the Verification Owner, promote them. If they are not, hire a Verification Owner from outside and pay them at the level the seat requires. Do not put a QA-band person in a CTO-band seat and expect the signature to carry the weight.

Tech Lead

Tech Lead is the in-pod technical authority. Decides the architecture. Reviews the PRs. Makes the call when two engineers disagree. The Tech Lead on a Scrum team is usually the most senior engineer, and the role is partly a career step and partly a coordination load.

The AI Orchestrator is the successor seat, and the overlap is close enough that most Tech Leads slide into it without much retraining. Both seats are the in-pod technical authority. Both seats review output. Both seats make architectural calls. The difference is what they are authorizing. A Tech Lead authorizes the merging of human-written code. An AI Orchestrator authorizes the shipping of machine-generated code against an acceptance contract the Intent Engineer authored. The judgment required is the same judgment, pointed at a different artifact.

The Tech Leads who struggle with the transition are the ones whose identity was tied to writing code themselves. The AI Orchestrator does not write code. The AI Orchestrator directs the generation, reviews the output, and intervenes when the direction is wrong. If a Tech Lead insists on writing the code themselves, they will be slower than the pod and worse than the machine, and the pod will route around them. The role is the same. The hands are off the keyboard. That is the adjustment.

Business Analyst

Business Analyst translated between the business and engineering. Wrote requirements. Wrote user stories. Wrote acceptance criteria. Ran the backlog grooming. Sat in the sprint planning. Translated when the product owner said something the engineers did not understand.

The Intent Engineer is the successor seat. The skill is substantially similar, and in practice the best Business Analysts make the fastest Intent Engineers, because they already think in outcome-shaped artifacts and they already know how to interrogate a vague request until it becomes a specific one. The Gherkin is a constraint they pick up in a couple of weeks. The WHY discipline is something they have been practicing for a decade.

The Business Analysts who do not make the transition are the ones whose value was tied to the volume of documentation they produced. An Intent Engineer produces one VOS per feature. A Business Analyst in a mature Scrum shop could produce a dozen documents per feature, and some of them measured their value by the page count. The Intent Engineer's page count is lower by design. If the Business Analyst cannot let go of the documentation volume, the role does not fit. If they can, it is the cleanest conversion in the gallery.

Project Manager

Project Manager ran the schedule. Ran the dependencies. Ran the risk log. Ran the status report up to the steering committee. In a PMO-heavy organization, the Project Manager was the person holding the line between what the engineers said was possible and what the business committed to.

The intent stream replaces the project. A stream is a standing budget for a standing outcome. It does not have a start date, an end date, a Gantt chart, a critical path, or a steering committee. The Project Manager's artifacts mostly have nothing to attach to.

This is the costume most organizations struggle with, because the PMO is usually a cost center with political weight, and the Project Managers are usually experienced operators who have survived multiple reorgs. The honest answer is that the Project Manager role does not have a direct successor in a CID organization. Some Project Managers convert to Intent Engineers, because the intent-writing skill maps to the requirements-writing skill. Some convert to Intent Portfolio Leads at the enterprise layer, because the cross-stream coordination job is real and does not disappear. Some do not convert.

The organization has to have that conversation honestly, rather than inventing a CID-labeled seat for the PMO to migrate into. That is the Forty-Person PMO anti-pattern, and the diagnostic is a CID org chart that has more coordinators than pods.

Release Manager, Change Manager, Environment Manager

These are grouped because the conversion is the same. The release is a signature on a VOS. The change is the VOS. The environment is managed by the harness. The three roles collapse into the Verification Owner's scope, and the artifacts the three roles used to produce are now either generated by the harness or not needed.

The person wearing one of these hats either takes the Verification Owner path or moves into an infrastructure role on the platform team. The Verification Owner path is rare, because usually only one of the three is strong enough to hold the seat. The infrastructure path is more common. The environment skill is still valuable. It is just not valuable at the application-pod boundary anymore.

Director of Delivery

Director of Delivery sat over a portfolio of projects and was accountable for the aggregate schedule. The role is essentially the Intent Portfolio Lead in CID, and the conversion is usually clean for the person and confusing for the organization. The individual is doing roughly the same cross-stream coordination work. The organization must stop asking for schedule reports against projects and start asking for outcome reports against streams. That transition is a finance problem more than a people problem, and Chapter 9 covers it.

4. What the costume gallery is not saying

I want to head off three readings of the gallery that I have heard, and that are wrong.

The first wrong reading is that these people are redundant. They are not. The organization still has the same work to do. The work is now routed differently, and the people can mostly be routed along with it. The headcount impact of a CID conversion is usually small in the first year, because the conversion is about reallocation rather than reduction. Headcount changes show up later, when the organization realizes it does not need to hire the next twenty coordinators it had budgeted for, because the coordination load is not growing with the delivery load. That is a hiring-plan change, not a layoff.

The second wrong reading is that titles do not matter. They do. HR systems run on titles. Compensation bands run on titles. The outside labor market runs on titles. A person who has been a VP of Engineering for five years and gets moved to a pod as an AI Orchestrator needs to keep a title that reflects their seniority in the market, or they will leave, and they should. The costume carries real weight for the person wearing it, even when it carries none for the pod. The organization's job is to keep the costume attached to the person while the pod runs on the pod role. Dual-designation is fine. "VP of Engineering, operating as AI Orchestrator on Pod Finaize" is a perfectly good business-card line. It preserves the market reality without pretending the pod is organized by title.

The third wrong reading is that the pod roles are the new costume, and we have just moved the theater. This one is the worry I take most seriously, because it is a real failure mode. If the Intent Engineer seat becomes a pay band that people lobby for, rather than a function that people perform, the methodology has grown its own HR structure, and the new structure is at risk of calcifying the same way the old one did. The defense is to keep the pod roles loosely coupled to pay and tightly coupled to work. At Alchemaize, the three of us have rotated through all three pod roles across the 35 applications. Nobody holds a pod-role title permanently. When the work calls for it, the roles shift. Pay bands are set by seniority and by the

Verification Owner standard, not by which pod role someone happens to be in this month. The pod role names the function. It does not name the person.

The honest test for whether pod roles have become costumes is whether people can move between them without a political fight. If a Verification Owner cannot take an AI Orchestrator seat on next month's pod because "the band doesn't match," the pod roles have calcified. If they can, the roles are still describing work, and the work is still being done by whoever is best placed to do it. Keep the test on the fridge.

5. How the conversion actually goes

Figure F17 · Symptoms, diagnosis, recovery. *The rough sequence by which a title-organized group converts to pod-organized work.*

In the field, the conversion from title-organized to pod-organized runs on a rough sequence. It takes about a quarter for a single engineering group of thirty to forty people. Longer if the organization fights it. Shorter if the Verification Owner is already in place and the first pod is already running.

Week one is the first pod. Three people. They are named into the three seats. The seats are explained. The first VOS is written in week two, shipped in week three, and the first Pipeline Review happens in week three. The pod is producing output before anyone has reorganized the chart.

Weeks four through eight are the second and third pods. By this point, the Intent Engineer from the first pod has trained the Intent Engineer from the second pod, and the Verification Owner is starting to write the harness standards. Three pods is nine seats. Nine seats pulled from a group of thirty-five people means the remaining twenty-six are still in their old titles. Most of them know something is happening, nobody knows quite what, and the org chart has not changed.

Weeks nine through twelve are the conversation about the old chart. This is the uncomfortable month. Scrum Masters learn there are no more ceremonies to facilitate. Engineering Managers learn the sprint is gone. QA Leads learn the best one of them is being promoted, and the others are being redeployed to the platform team or to an Intent Engineer track. Project Managers learn about streams, and the good ones see the Intent Portfolio Lead seat and start reading about it.

The honest organizations have the conversations honestly. The dishonest ones invent CID-labeled seats for everyone and tell themselves nothing has changed. I have seen both. The honest version produces a working organization inside a quarter. The dishonest version produces a costume organization, which is the Costume CID anti-pattern writ large, and it reverts to Scrum muscle memory inside two quarters every time.

Month four through month six is the third wave of pods and the enterprise layer. The Intent Portfolio Lead and the Verification Officer seats are named. The finance function starts talking about streams instead of projects, usually badly at first. The HR function starts adjusting title bands, usually badly at first. The old chart is still visible, but it is visibly being replaced, and the pods are shipping faster than the costume organization can keep up with.

By month seven, a company that took the conversion seriously has three to five pods running, twelve to fifteen people in pod seats, an enterprise layer with two seats, and the balance of the old engineering group either redeployed to platform, converted to pod seats, or departed.

The departures are real and worth naming. Not everyone wants to work this way. A senior engineer whose identity is writing code alone is going to find CID uncomfortable. A coordinator whose identity is convening meetings is going to find CID unpleasant. A manager whose identity is directing a team by a sprint cadence is going to find the cadence gone. Some of those people adapt. Some leave. The leaving is usually clean, and in my experience the ones who leave land well. They were not wrong about what they wanted out of a job. They just wanted a different job than CID offered.

The net headcount change in the engineering organization after a full conversion is usually small in the first year, as mentioned. In the second year, the organization tends to grow its pod count and shrink its coordinator count, and the ratio of people-doing-the-work to people-coordinating-the-work shifts hard. That ratio is the real productivity gain. It shows up in the P&L slowly, because it takes a year or two of not hiring the coordinators you would have hired under the old model before the finance function notices. When they notice, it is a reliably large number.

6. The CTO and the CIO

The CTO job does not change. In a pre-CID organization, the CTO is accountable for technology strategy, technical risk, architectural direction, and the standard of engineering the organization holds itself to. In a post-CID organization, the CTO is accountable for the same four things. The artifacts have moved. The headcount mix has shifted. The verification architecture is new. The job is the same.

If anything, the CTO's job gets easier, because the verification signature is a real signal rather than a ceremonial one, and the CTO can read the first-pass verification rate and the four metrics and know what is actually true. The CTO who tries to hold onto the role of technical-decider-of-last-resort will be overworked. The CTO who trusts the Verification Owner signature and reads the dashboards will have time to think, which is what the role was always supposed to be for.

The CIO job, similarly, does not change. The CIO runs the technology function of a business, and is accountable to the board for the business outcomes that function delivers. CID changes how the function operates internally. It does not change the outcome accountability. A CIO running a CID organization has a quieter delivery organization, a higher first-pass rate, a lower defect rate, and a budget that shifts from coordinator headcount to pod-and-platform investment. The board conversation is simpler, not harder. The role is the same.

The reason to name this is that CIOs and CTOs sometimes fear a CID adoption as a threat to their own seat. It is not. The threat is to the middle of the org chart, the coordinator layer between the executive and the doing. The top of the chart is fine. The bottom of the chart is fine, and busier with interesting work. The middle is where the costume question gets asked hardest, and the middle is where the work of answering it must happen.

7. A dry aside about HR systems

A brief operational note, because it comes up in every engagement, and I have watched two conversions nearly fail on it.

HR systems are not designed to represent pod-role assignments. They are designed to represent reporting lines, pay bands, titles, cost centers, and location codes. A pod is a cross-functional assignment that does not fit neatly into any of those fields. The Verification Owner's reporting line is out-of-delivery, which the HR system will try to represent as a matrix, which is a word HR systems use when they have given up. The Intent Engineer's title is usually a legacy title from before the conversion, which the HR system will reconcile against a pod-role field that does not exist.

The solution is not to rebuild the HR system. The solution is to keep the HR system doing what it does (payroll, bands, benefits, compliance) and to keep the pod-role assignment somewhere else. A shared spreadsheet. A

Confluence page. A field in the VOS system. An ELCID governance artifact. Name it, version it, and do not ask HR to be the source of truth for it. HR can be the source of truth for the costume. The pod can be the source of truth for the role. Both are correct. They answer different questions.

I mention this because two of the three conversions I saw stall in 2025 stalled on this exact fight. The HR director insisted pod roles had to live in the HR system. The HR system could not represent them. The conversion stopped while a tooling project was scoped. The tooling project took a quarter. The conversion waited. The momentum was lost. Do not do this. Keep the two systems separate and move on.

8. The costume test

There is a test I run in engagements when I am trying to figure out whether a CID conversion is real or cosmetic. It takes ten minutes. I ask five questions.

First: can you tell me, today, who is the Intent Engineer on the pod shipping the most VOSes this month. Name the person. If the answer is a title, a team, a function, or a department, the pod does not have an Intent Engineer. It has a committee with a label on it.

Second: what is the Verification Owner's reporting line, and is it in delivery or out of delivery. If the answer involves the word "matrix," the reporting line is effectively in delivery. If the answer is "the CTO" or "the Verification Officer," the role is structurally intact.

Third: how many standing meetings does the pod have on its weekly calendar. One is right. Zero is better. Two means the organization has invented a ceremony. Three or more means the conversion is cosmetic.

Fourth: when was the last time a VOS was rolled back at the artifact level, and how long did the rollback take. If nobody remembers, the rollback path has not been exercised, and the Verification Owner has not earned their signature. If the answer is "within minutes," the harness is real.

Fifth: if the Intent Engineer on this pod quit tomorrow, who takes the seat. If the answer is "nobody on this team could," the pod has a role-supply problem. If the answer names two or three people on adjacent pods who could step in, the roles are genuinely shared, and the organization has built a bench.

Three out of five right means the conversion is real and healthy. Two out of five means the conversion is in progress, and the costumes are starting to outrank the roles, and someone needs to call it out this quarter. One out of five means the organization has done a rebrand rather than a conversion. Zero out of five means the org chart has CID words on it and nothing else has changed, and the Costume CID diagnosis applies at the whole-organization level, not just at the ceremony level.

I have run this test maybe thirty times. The distribution is bimodal. Most organizations are at three out of five, or at zero out of five. Almost nobody is in the middle. The conversion either takes or it does not. The middle cases I have seen were usually organizations mid-quarter, on their way to one of the two stable states, and the direction they moved depended almost entirely on whether the CEO or the CTO was willing to have the uncomfortable conversation in the middle of month three. If they had it, the conversion took. If they deferred it, the conversion drifted, and the drift was usually fatal inside six months.

9. The Scrum Master tell

The single most predictive signal of a failing CID conversion is the preservation of the Scrum Master title on the post-conversion org chart. The person is fine. The title is the tell.

If, six months in, the chart still has "Scrum Master" on it as a standing role, the organization did not convert. It printed new business cards.

The title is specifically the signal, because every other title in the gallery has a plausible post-conversion role to land in. Scrum Master does not. Scrum Master exists to run the ceremonies that CID removes. A Scrum

Master title on a CID chart is the organizational equivalent of a ticket booth at a toll bridge that was replaced by a transponder a year ago. The booth is still there because nobody tore it down. The tickets have stopped selling.

The booth costs money. The person in the booth is bored. The transponder is doing the work. Tear the booth down.

10. What to tell the person in the hat

One conversation remains.

The chapter is diagnostic. The diagnosis is unsentimental. The titles are costumes, and the roles are the work. All of that is correct, and none of it makes the conversation with the person in the hat any easier.

The person in the hat has spent a career earning the title. The title is real to them. They built a practice. They built a network. They built a professional identity around the work the title used to describe. The work has moved, and nobody asked them before it did. The book is now telling them their title is a costume. That is not a thing the book has the standing to say lightly.

I do not say it lightly. I have been on the other side of it. Some of the titles in my own career, including platform architect, VP of Innovation, and SVP of Consulting and Services, cover work that has changed shape in the last five years. The parts of my own practice that I thought were the work turned out to be the scaffolding around the work. I have had to reconvert more than once. It is not fun. It is also not the end of the career.

The people I know who handled the reconversion well did three things.

They named the costume. They did not pretend the title still meant what it used to mean. They said, out loud, "this title used to cover X. X is now done by a pod role. I need to figure out where I am going next." Naming it is half the work.

The other half is the decision, which follows. They picked a pod role to learn, one rather than all three, and the one closest to their existing skill. A

QA Lead who picked Verification Owner. A Business Analyst who picked Intent Engineer. A Tech Lead who picked AI Orchestrator. The overlap between the old title's craft and the new role's craft is usually large, and the conversion is measured in weeks rather than years. The people who tried to learn all three roles at once did worse than the people who picked one and went deep.

They kept the market-facing title. The costume did not come off until the market rewarded them for doing the pod-role work, which took about a year. In the meantime, they were "VP of Engineering, currently running the AI Orchestrator seat on Pod X" on their LinkedIn and at their professional associations, and nobody asked too many questions. The transition is allowed to be gradual. The methodology does not require the costume to come off the same week the role changes.

The conversion is uncomfortable. It is not a demotion. It is a relocation of craft into a system where the craft is measured by its outputs rather than by the meetings it runs. Most of the people I have watched make the move have ended up doing work they find more interesting than the work they left. Some have not. Both outcomes are real.

11. Close

Titles are hats. The work is the thing. When the work moves, the hat does not move with it, and the organization ends up with a rack of hats that no longer correspond to any seat where work is being done. Naming that honestly is the first step to doing something about it.

CID does not abolish the hat rack. It reorganizes the work underneath it. The three pod roles describe the work. They are not themselves titles, and they should not become titles, because the moment they do, the methodology has grown its own costume rack and we are back where we started. Treat the roles as functions and the titles as market-facing labels, and keep the two clearly separated in your own head before you ask the organization to do it.

Team structure is a variable the organization gets to set, not a ceremony it has to keep running. That claim is not original to this book. Team Topologies made the same argument years before CID arrived, organizing the question around four patterns: stream-aligned, platform, enabling, and complicated-subsystem, with cognitive load as the binding constraint.

CID inherits the posture. A pod is a stream-aligned team. The Verification Owner's scope sits closer to an enabling function. The platform team is where the former environment and release managers land.

The costumes are the leftover shapes from an era that treated team design as a ceremony output. When team design becomes a variable again, the costumes come off.

Look at the rack. Name what is on it. Take the empty hats down.

That is most of the work.

The pod ships. The costume is optional.

Notes

1. Ken Schwaber and Jeff Sutherland. *The Scrum Guide,* 2020 edition. scrum.org/resources/scrum-guide (accessed 2026-04-22).
2. Scaled Agile, Inc. *SAFe 6.0 for Lean Enterprises.* Scaled Agile, 2023. scaledagile.com/safe-big-picture (accessed 2026-04-22).
3. Matthew Skelton and Manuel Pais. *Team Topologies: Organizing Business and Technology Teams for Fast Flow.* IT Revolution, 2019.

CHAPTER 13: THE ADOPTION PATH

The hardest chapter in a methodology book is the one that tells the reader what to do on Monday morning. The rest of the book is argument. This chapter is an assignment.

I want to open with the assignment itself, stated plainly, so that everything after it is detail. The assignment is to pick one team, ship one VOS, and let the next VOS be easier than the first. That is the whole path. Chapter 14 closes the book on that sentence. Chapter 13 is the twenty-six weeks between deciding to start and the moment you're no longer doing this as an experiment.

A reader who has read the previous twelve chapters can be forgiven for expecting a transformation program here. Most books of this kind, at this point, produce one. A six-stage maturity model, a role-based enablement track, a wall of RACIs. I'm going to decline to produce that. The reason is not stylistic. It's that the methodology we've described in this book does not require a transformation program, and the organizations I've watched try to adopt it through a transformation program have had the worst outcomes of any organizations I've watched.

CID is adopted the way CID is practiced. One pod. One stream. One VOS. The enterprise layer comes online when the pod has something worth scaling, not before. If that sounds deflating to a reader who was hoping for a five-year roadmap, the deflation is the point. A methodology whose main promise is that it retires coordination overhead cannot be adopted by a coordination overhead.

What follows is four phases. The first four weeks are concrete enough that a reader should be able to execute them with nothing more than this chapter, Appendix B (the worked VOS), and Appendix C (the self-assessment). The next twenty weeks have more variance because the shape depends on what the first four weeks teach the organization. Phase four is open-ended, because scaling CID is something organizations do continuously rather than finish.

1. Phase 1: one pod, one stream (weeks 1 to 4)

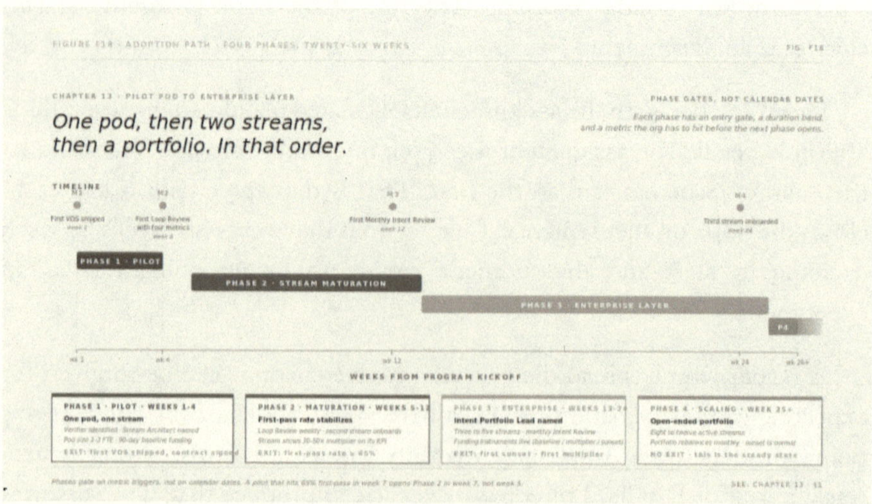

Figure F18 · Adoption path, 4 phases. *Phase 1 (weeks 1 to 4), phase 2 (weeks 5 to 12), phase 3 (weeks 13 to 24), phase 4 (month 7 and beyond).*

The first phase has three deliverables and nothing else. Identify the pod. Identify the stream. Ship the first VOS by the end of week four.

That sentence contains the chapter's whole instruction for the first month. Everything else here is elaboration.

Identify the pod

A CID pod is three people or a two-plus-fractional arrangement where one role is partial. Intent Engineer, AI Orchestrator, Verification Owner. The roles are described in Chapter 4 and Chapter 6. What I want to say here is about the people, not the roles.

The Intent Engineer should be someone who already writes requirements that engineers treat as usable. At most organizations this is a product manager or a senior business analyst. Occasionally it's a former customer-success lead who has spent years translating what customers actually want into language a delivery team can act on. Casey is the prototype. There are more people like Casey than the industry currently lets do this work.

The AI Orchestrator should be a senior engineer who is comfortable generating, reading, and redirecting code produced by a model. The seniority matters. A mid-level engineer will do the mechanical parts of the role well and miss the judgment calls. The judgment is the role. If your candidate has spent their career demanding green builds and has lately been skeptical of AI coding tools, that skepticism is a qualification, not a disqualification. The role needs someone who will say the generated code is wrong when it is wrong.

The Verification Owner should be someone senior enough to say *no, not yet* to a shipping pressure they did not create. Most organizations have this person and call them a QA lead, a senior SDET, or a principal engineer on the platform team. The critical property is organizational standing. A Verification Owner who reports to the delivery chain will eventually stop saying no. Chapter 6 explained why. I'm going to repeat the mechanics here because the first pod stand-up is where this usually breaks. The Verification Owner does not report to whoever is pushing the feature.

One practical consequence of the three-role structure is that the pod can be two-plus-fractional. A product manager full-time on Intent Engineering, a senior engineer full-time on Orchestration, and a principal engineer who spends half their week on Verification for this pod and half on something else. That works. What does not work is the four-role structure with a designated Scrum Master. The coordination is in the roles. Adding a coordinator re-creates the ceremony burden the methodology is designed to retire.

Identify the stream

A stream is a funded line of intent aimed at one outcome. Chapter 9 described streams at the enterprise layer. At the team layer, in week one, the stream is simpler. Pick one product area. Pick one outcome the product area is supposed to produce. Pick one KPI the outcome is measured against.

The examples that work best in first pods have three properties. The outcome is something the business already cares about measuring, so the team is not inventing a new metric alongside a new methodology. The product area is small enough to hold in one head, so the Intent Engineer can write a VOS without paging in the entire organization's data model. The KPI moves on a weekly horizon rather than a quarterly one, so the pod can see whether the work shipped in week four is having the intended effect by week six.

A stream I have seen work: *reduce the support-ticket volume on the billing reconciliation flow.* The outcome is billing reconciliation, which the finance team already watches. The product area is one screen, one workflow, and three backend modules. The KPI is tickets per week, which moves on a weekly horizon.

A stream I have watched fail: *improve the platform.* This is not a stream. It's a wish. The first VOS against it will drift into architecture discussion, the acceptance contract will be impossible to write, and the pod will abandon CID by week three because the methodology didn't protect them from an unbounded goal. Protect the pod from unbounded goals.

Write the first VOS by week two

The Intent Engineer authors the first VOS in week one and submits it to the Verification Owner for acceptance-contract review in week two. The first VOS will be rejected. This is expected. Casey's first VOS was rejected on three of four acceptance contracts, and two of those rejections were correct. The first pod should plan for the same outcome and treat the rejection as training data rather than as failure.

The appendix that sits behind this chapter, Appendix B, walks through a worked VOS end to end. The pod should read Appendix B before writing their first one. Appendix B is not a template. It's a reference case, and the pod's first VOS will look different because the stream is different. What transfers from Appendix B is the shape: WHY, WHAT as an executable acceptance contract, HOW at the design-decision level, CONTEXT as four to fifteen named files, OUTCOME as a measurable movement in the stream's KPI.

Run the first Pipeline Review by week four

The Pipeline Review is described in Chapter 7. Thirty minutes, weekly, three people. The pod will want to make it longer the first time. Do not make it longer. The thirty-minute constraint is part of what the methodology is testing.

Four numbers at the Pipeline Review. Intent cycle time. First-pass verification rate. Outcome KPI delta. Cost per shipped intent. These four numbers and no others. Chapter 10 of this book covers the metrics in full and lists the exact definitions; the pod should use those definitions rather than close approximations.

At the end of week four the pod should have shipped between one and three VOSes, run four Pipeline Reviews, and established a baseline on the four metrics. Nothing else is expected. The enterprise layer does not exist yet. The Portfolio of Intents does not exist yet. The Monthly Intent Review does not exist yet. The temptation to build any of those in parallel should be resisted. The first pod's job is to prove the pipeline works on the stream they picked.

One quiet fact about phase 1. Most of the organizations I've watched succeed at adoption did not start by telling anyone outside the pod what they were doing. The pod ran for four weeks, produced evidence, and then the evidence was what the broader organization reacted to. The organizations that announced they were adopting CID at a town-hall meeting before week one had already converted the thing into a change-management exercise, which is the wrong category.

2. Phase 2: stream maturation (weeks 5 to 12)

The second phase is where the first pod stops being an experiment and starts producing a cleaner stream of work than the pre-CID baseline. Eight weeks, in which the four metrics move in visible directions.

What the pod does in weeks 5 through 12 is tune the acceptance contracts to the specific risks of the stream they picked. The first VOS's acceptance contract was a rough one. By week eight it should be a tight one. The Verification Owner is the person tuning it, in collaboration with the Intent Engineer, and the tuning is specific to what has been failing in verification so far.

First-pass verification rate is the number to watch. At week four it will be somewhere between 30 and 50 percent, which is normal. At week eight it should be in the 50 to 70 percent band. If it isn't, the pod has a diagnosable problem. Either the Intent Engineer is writing VOSes that are too large and the acceptance contracts are missing cases the scope implied, or the AI Orchestrator is not resetting context between tasks and the generation is drifting, or the Verification Owner is writing contracts at a rigor the stream doesn't actually need, which produces false rejections that look like low first-pass rates. Chapter 6 covers each of those diagnostics and the corresponding fixes.

Deployment of the verification infrastructure is the other thing that happens in phase 2. The acceptance contracts the pod has been writing by hand become deterministic verifiers wired into the pod's continuous integration. Blocking-finding enforcement comes online. The Verification Owner's veto is no longer a conversation; it's a build status. This is a mechanical step and the training material in `training/04-verification-owner.md` walks through it.

If the stream is in a regulated space, phase 2 is also when the compliance extensions described in Chapter 11 come online. HIPAA, SOC 2, PCI-DSS, FedRAMP, CMMC. These are contract templates that sit inside the acceptance-contract layer and block non-compliant generation at generation time. Chapter 11 made the argument that regulated industries are the strongest candidates for CID, not the weakest. Phase 2 is where that argument becomes operational. The compliance extensions do not add ceremony. They add verification rules. The pod notices them the way a developer notices a lint rule, which is to say barely.

By the end of phase 2 the pod has shipped between eight and twenty VOSes, has first-pass verification rates climbing into the 50 to 70 percent band, has cycle times stabilizing somewhere between one and three days per VOS, and is producing cleaner output than it produced in its pre-CID baseline. That last clause is the point. Phase 2 is the phase that proves, to the pod and to anyone watching the pod, that the methodology produces better work, not just faster work. If phase 2 does not produce cleaner output, phase 3 should wait.

3. Phase 3: the enterprise layer (weeks 13 to 24)

Phase 3 is where the book's Part III comes online. Portfolio of Intents. Lean Intent Funding. The Intent Portfolio Lead and the Verification Officer appointed. The first Monthly Intent Review, ninety minutes, replacing the quarter's worth of PI Planning the organization used to do.

I want to be specific about what brings this phase on, because the sequencing matters. Phase 3 is not triggered by a calendar. It's triggered by the first pod having something worth scaling and the organization having a second or third stream that needs funding. An organization that runs one successful pod for six months and does not add a second does not need phase 3. Phase 3 is the architecture that handles two or more streams under one funding envelope.

The Intent Portfolio Lead is the person who decides which streams get funded, at what levels, for the coming month. This role replaces the project-selection committee at most organizations, or the product council, or whatever the legacy instrument is for saying what gets built. The replacement is real, not aspirational. A Portfolio of Intents is a working document that lists the current funded streams, their budgets, their outcomes, and their current status. It is updated monthly. It is not a PowerPoint deck.

The Verification Officer is the enterprise counterpart of the pod-level Verification Owner. This is the person who owns the verification function across all streams, sets the standards for acceptance contracts at the organization level, and reports independently to whoever governs quality at the enterprise. In a company with a CISO, the Verification Officer often reports into the CISO. In a company without one, the role reports into the CTO but with explicit veto authority. The reporting line matters, again, because a Verification Officer who reports to the delivery chain will eventually stop saying no.

The Monthly Intent Review is ninety minutes. One meeting. All streams represented by their pod leads, the Intent Portfolio Lead facilitating, the Verification Officer in attendance with veto authority on any stream that is not shipping against its acceptance contracts. Each stream gets ten minutes. Four metrics, one outcome update, one decision if a decision is needed. The review is not a status meeting. Status is in the metric dashboards, which anyone can pull up. The review is for funding decisions and cross-stream tradeoffs.

If the organization is running more than eight streams, phase 3 also stands up Stream Architects. The Stream Architect is described in Chapter 9, and the role handles cross-stream technical coherence. Architecture decisions that affect more than one stream run through the Stream Architect. At fewer than eight streams, the coherence can be handled by the pod-level Orchestrators in an hour every two weeks. Beyond eight, the coherence needs a person.

By the end of week 24 the organization has a functioning enterprise layer. The streams are funded monthly. The metrics are visible. The Portfolio of Intents is a living document. The Verification function is independent. No PMO. No SAFe release train. No PI Planning. No Scrum of Scrums.

4. Phase 4: scaling (month 7 and beyond)

Phase 4 is indefinite. The question it answers is how the organization grows the methodology without turning the methodology into the thing it replaces.

The binding constraint on phase 4 is role supply. An organization that wants to run ten streams needs ten Intent Engineers, ten AI Orchestrators, and ten Verification Owners, or the fractional equivalents. Those people have to come from somewhere. The organization's job in phase 4 is to identify, train, and retain the people who fill those roles. That is the job. It is not a coordination job. It is a hiring, developing, and retaining job.

Scaling CID is horizontal. You add streams. You do not add coordination layers between the streams. The reason is simple. The methodology's design assumption is that the expensive step is intent and verification, and the cheap step is generation. Coordination layers are a solution to a different assumption, namely that generation is the bottleneck and multiple generators must be synchronized against each other to ship. That assumption is not true under CID. The streams are independently funded, independently shipped, and independently verified. A coordination layer between them would be paying coordination costs to manage an independence the methodology guarantees.

The sequencing of roles here borrows its vocabulary from Matthew Skelton and Manuel Pais's *Team Topologies*, which names four patterns for organizing around flow: stream-aligned, platform, enabling, and complicated-subsystem [1]. The CID pod is a stream-aligned team by another name. The platform group that handles environments and shared infrastructure maps onto *Team Topologies*' platform pattern. The Verification Guild and the Center of Excellence for Intent Engineering, if the organization builds them, are enabling teams; they supply patterns rather than gate delivery. A complicated-subsystem team appears only when a specific domain (a pricing engine, a billing reconciliation module, a compliance kernel) has enough intrinsic complexity to warrant its own standing pod. Most organizations never need that fourth pattern in the first twenty-four weeks; when they do, the methodology accommodates it without changing the pod shape.

What does happen in phase 4, mechanically, is a Center of Excellence for Intent Engineering, if the organization is large enough to need one. The Center of Excellence is not a team that writes VOSes for other teams. It's a training, hiring, and standards function. It runs the organization's Intent Engineer bench. It writes the house style for acceptance contracts. It maintains the shared library of contract templates that pods can pull from. It does not gate. It supplies.

The same pattern holds for Verification. A Verification Guild, cross-stream, meets monthly. Shares what each stream has been finding in verification. Updates the shared contract templates. Does not gate any individual pod. Supplies patterns that the pods adopt or don't.

If the organization is large enough to need a PMO to coordinate thirty teams, the right answer in phase 4 is not to add a PMO over the streams. It's to ask whether thirty independent streams should all be independent, or whether some of them should merge into larger streams with bigger envelopes. The answer is almost always the latter. A stream is not a team; it is a funded intent. One stream can contain multiple pods if the intent is large enough. Scaling is done by re-scoping, not by coordinating.

The last piece of phase 4 is the retention signal. Intent Engineers and AI Orchestrators and Verification Owners who have spent six months in a CID pod do not want to go back to Scrum. The pods become retention magnets.

5. Five CIO objections, with the data response

Objection	The data response
"We spent millions on Agile transformation. We can't throw it away."	The framework is what you retire. The people convert: Scrum Masters who understand flow become Intent Engineers, senior engineers become AI Orchestrators, QA leads become Verification Owners.
"AI-generated code is risky."	Unverified AI-generated code is risky. Verified AI-generated code is safer than human-written code. Measured rate across the work this book covers: under three defects per 1000 lines against a McConnell-cited industry range of 15-50.
"Our engineers will revolt."	The engineers who try it do not revolt. Pattern: week one disorienting, week four first confidence, week twelve realizing they ship better code than before. Engineers who have been in a CID pod for six months do not want to go back.
"We're too big."	The methodology scales by adding streams, not coordination. A hundred-engineer organization running twenty independent streams ships more than the same hundred coordinated through SAFe.
"We can't find Intent Engineers."	The role is learnable. Casey is the proof (Chapter 8). The skill is clarity under pressure, which your organization already has in product managers, senior business analysts, and customer-success principals. Hire for the disposition; train the AI fluency in three to six weeks.

Table T8 · Five CIO objections. A sixth objection ("we're in the middle of a SAFe rollout; can we finish it first?") is addressed in the prose that follows.

Every adoption conversation I've had with a CIO has hit five walls. They are the same five walls every time. I'm listing them here with the answers, compressed to the shape a CIO can use directly.

"We spent millions on our Agile transformation. We can't throw it away." You're not throwing away the people. The Scrum Masters who were any good at the job were good because they understood flow and impediments, and they'll become good Intent Engineers. The senior engineers who ran architecture decisions become AI Orchestrators. The QA leads become Verification Owners. The framework is what you're retiring. The people are the asset and they stay. The sunk cost is real. Continuing to deepen it is worse.

"AI-generated code is risky." Unverified AI-generated code is risky. Verified AI-generated code is safer than human-written code. The measured defect rate from the work covered by this book is under three bugs per thousand lines of code, against an industry baseline around fifteen. The verification gate is the thing that makes the speed safe. Chapter 6 describes the gate. The risk argument is correct about the unverified case and wrong about the verified case, and the methodology is entirely about the verified case.

"Our engineers will revolt." The engineers who try it do not revolt. They want more of it. The pattern I have watched in every pod I have watched is the same. Week one is disorienting. Week four is the first real confidence. Week twelve is the moment the engineer in the AI Orchestrator chair realizes they are shipping better code than they used to, and they stop having opinions about the methodology and start having opinions about the next stream they want to run. Measure the retention signal. Engineers who have been in a CID pod for six months do not want to go back.

"We're too big." The methodology scales by adding streams, not by adding coordination. Large deployments on CATALYST today run more streams in parallel than a SAFe release train could coordinate in a quarter, and do so without the PI Planning week the release train required. The size is the opportunity, not the obstacle. A hundred-engineer organization that runs twenty streams independently ships more work than the same hundred engineers coordinated through SAFe, because the coordination cost in SAFe is roughly thirty percent of the capacity and CID does not spend it.

"We can't find Intent Engineers." The role is learnable. Casey is the proof, documented at length in Chapter 8. The skill of writing a VOS is the skill of clarity under pressure, which your organization already has people who can do. They are currently titled product managers, senior business analysts, or customer-success principals, and some of them are excellent and underused. Hire for the clarity skill. Train the AI fluency on the job. The training takes three to six weeks for a candidate with the underlying disposition and does not take at all for a candidate without it. Hire for the disposition.

There is a sixth objection that CIOs rarely name in the first meeting and often name in the second. *We're in the middle of a SAFe rollout; can we finish it first?* The honest answer is no. The sunk cost will compound if you continue. SAFe, per `scaledagile.com`, prescribes an implementation roadmap that moves an enterprise through training, Agile Release Train launches, program-increment cadences, and portfolio-level Lean Portfolio Management in a sequence that typically takes twelve to twenty-four months to stabilize [2]. That roadmap is a rollout of coordination infrastructure designed for a constraint that does not bind your organization anymore. Finishing it makes the eventual unwind more expensive, not less. Stop it. The dollars already spent are gone. The dollars about to be spent are still yours. Spend them on the methodology that fits the constraint you actually have.

6. Casey co-lead: what the phases feel like from inside

Casey Robinson, Intent Engineer.

Week one is disorienting, and I'm going to say that openly because it's the one thing I wish somebody had said to me before I started.

The disorientation isn't about the tools, because the tools are fine. The disorientation is that a CID pod runs differently than every other software-delivery setup I'd been in, and the rhythm I thought I'd bring from twenty years of Agile work didn't help and in some cases actively hurt. I kept trying to schedule a standup, I kept trying to build a backlog, I kept trying to set a sprint boundary, and the pod I was in didn't need any of those things. The hour I spent trying to build them was an hour I didn't spend writing the first VOS.

The first VOS is the week-one deliverable, so write it, and try not to do anything else first.

Week two is calibration. The pod has reviewed the acceptance contracts on the first VOS and some of them came back with feedback. Some of the feedback is about real gaps in the contract. Some of it is about the pod finding its shared language. The work of week two is learning which is which, and learning the specific flavor of precision your stream needs. A billing-reconciliation stream needs a different precision than an onboarding-flow stream. You don't know what yours needs yet. You're learning it from the feedback.

Week three is when the writing gets faster. The first VOS took me a while, the second less, and by the fifth I was writing them in a fraction of the time it had taken to write the first. I'm not a fast writer at all. The writing gets faster because the shape gets familiar, and the shape does what shapes always do once you know them, which is it stops taking up brain to produce.

Week four is the first Pipeline Review with real metrics, and the numbers aren't going to be impressive. Intent cycle time is maybe three days. First-pass verification rate is maybe 40 percent. You've shipped two VOSes. Don't be impressed, and don't be embarrassed. The numbers at week four are the baseline, and the only thing that matters is whether they move in the right direction by week eight.

Weeks five through twelve are where the methodology starts doing the thing it was designed to do, which is produce cleaner work than you produced before. This is the phase where I realized, sitting at my desk in my office in Claremore, that the VOSes I was writing now had fewer defects reaching production than the feature specifications I'd written in my AWS job had produced when passed through a human engineering team. That was the week I stopped being skeptical. If you're an Intent Engineer running your first pod, that moment is somewhere in your weeks five through twelve. You'll know it when it arrives. Don't announce it. Just keep writing VOSes.

Week thirteen is when the enterprise layer starts to arrive, if the organization is building one. My first Monthly Intent Review felt strange. Ninety minutes, three streams, all four metrics visible on a dashboard anyone could pull up, and the decisions that used to take three meetings took one. I kept expecting somebody to call for a breakout session. Nobody did. The meeting ended on time.

Week twenty-four isn't a milestone, it's just a date. What actually happens between week twelve and week twenty-four is that the pod becomes routine. The methodology stops being the thing you're practicing and starts being the thing you're working inside, and this is the phase where newer pods spun up elsewhere in the organization will look to you for what to do, and you find yourself explaining things you didn't know you knew. That's the phase I'm in now.

Month seven forward is, for me, the phase where the thing I do stopped being an experiment. I ship VOSes the way a product manager ships features, which is to say routinely, without drama, without ceremony, and with a quiet understanding that the work is real.

Here's the one thing I'd add for the reader who is about to try this. The four phases aren't four difficulty levels, they're four stations in a work routine. The hardest phase, if you're doing it right, is phase one, and it's hard because you're learning to write a VOS. Each phase after that gets more familiar, not harder. If a later phase feels harder than an earlier one, something is probably off, and the thing that's usually off is that the organization has added coordination where none was needed. When you feel that, trust what the phase is telling you.

7. Distributed by default

A CID pod does not need a room.

I want to say that directly because the opposite claim is still being made in a lot of enterprise engineering shops, and the claim is always packaged as a methodological argument when it's usually a preference dressed up as a method. *We need the team in the room for the standup. We need the whiteboard for sprint planning. PI Planning has to happen in person.* I've heard every version of this, and I've run meetings that used every version, and I've watched the version collapse under its own weight when the bottleneck moved.

The old ceremonies were scaffolding for a workflow that was typing-bound. If twelve humans have to coordinate their typing across a two-week sprint, the whiteboard helps, the standup helps, the room helps. Co-location was never a truth about software. It was a truth about coordination costs inside a specific bottleneck. When the bottleneck moves, the room loses its function.

CID's artifacts travel differently than Scrum's artifacts. A written VOS travels over the wire at the speed of Git. An acceptance contract is an executable file. A context bundle is a directory of references. A verification report is a CI run. A pipeline review is thirty minutes, three people, one video call, four numbers on a screen anyone can share. None of these artifacts require a whiteboard, a room, a zip code. They require the three people in the pod to be paying attention, which is independent of where the three people are sitting.

Alchemaize itself is evidence. Three founders, three states. Brentwood, Tennessee. Claremore, Oklahoma. College Station, Texas. Every VOS described in this book, every application in the hundred-day sprint, every piece of the methodology, was shipped without the three of us sharing a zip code. We have never shared an office. We never will. The company was founded distributed and is staying that way because the methodology does not benefit from co-location and the people on the team don't want to move.

If the organization adopting CID is currently running a Return-to-Office policy, that policy is worth examining in light of the methodology. The RTO argument rests, almost entirely, on the value of in-person coordination for a coordination-heavy workflow. CID is a coordination-light workflow. The arithmetic changes. I am not going to tell any organization what its RTO policy should be, because the policy depends on a lot of things that aren't in this book. But I will say the methodological argument for RTO does not survive the methodology. The pull toward co-location, when it's there, is usually a signal that the old methodology has not yet been put down.

Casey wants to add one paragraph to this section.

One paragraph from Casey. I've been at my desk in my office in Claremore for every VOS I've shipped. David has been in his home office in Brentwood. Glenn has been in his home office in College Station. None of us have been in a room together in the working period this book covers, and the book got written too. The artifacts travel fine. The methodology travels fine. When an organization exploring CID adoption asks me what the virtual-versus-in-person tradeoff is, my honest answer is I don't know, because it never came up for us. The organizations I've watched struggle have been the ones still running Scrum ceremonies in conference rooms while asking the methodology to deliver post-Scrum outcomes. That's not a virtual-versus-in-person problem. That's a methodology-hasn't-been-put-down problem.

I appreciate David and Glenn giving me the space to share what these phases look like from the inside. Now let me hand it back over to David to bring the chapter home.

8. The chapter's quiet commitment

One quarter. Measurable results in four weeks. One pod, one stream, one VOS at a time. The path is not a transformation program. It is a path.

The reader who has finished this chapter and is ready to take the first step should turn to Appendix C. The readiness self-assessment is roughly thirty questions across five sections, twenty minutes alone, one hour with three to five colleagues. It surfaces the readiness gap most likely to sink the first pod, naming which of the five sections (pod roles, stream selection, verification readiness, funding and governance, or anti-pattern check) is the weakest link before the work begins. That diagnosis is the last thing this book owes the reader.

Pick one team. Ship one VOS. The next one will be easier.

Field Note from the Test Pilot

> *Casey Robinson, Intent Engineer*
>
> **What I wish I'd known in week one, and what I think was useful I didn't know.**
>
> I'm going to list both, because the list is more honest that way.
>
> **What I wish I'd known.**
>
> The context-bundle discipline. In week one I was including too much. I was pointing the AI at everything because I didn't know what it needed. The AI does not need everything. It needs the four to fifteen files that are directly relevant to the VOS, and my AI tool helped me identify which ones those were once I could describe the feature area in plain language. I learned this by week three. I could have learned it on day one if somebody had told me.
>
> The Gherkin form. I had written user stories before. I had written acceptance criteria. I had not written Gherkin. Gherkin is not acceptance criteria with colons added. It is an executable description of behavior. The *Given / When / Then* structure carries real weight in a way I did not appreciate until I saw a malformed Gherkin block fail to bind to any actual verification. If I had known this on day one I would have spent week one reading Gherkin examples instead of writing bad ones.
>
> The rhythm of the Pipeline Review. Thirty minutes. Four numbers. Three people. I tried to make the first Pipeline Review

longer because I had a lot to say, and I had a lot to say because I hadn't been in a thirty-minute meeting in a decade and I didn't trust the format. The format is right. Thirty minutes forces the discussion onto the four numbers and off the feelings about the four numbers. The feelings about the four numbers are not what the review is for.

That the feedback rate on my first VOS was a training signal, not a grade. On VOS number one, a chunk of the acceptance contracts came back for rework, and I took it personally for about forty-five minutes before I caught myself. The feedback isn't a grade, it's the pod doing exactly the job the methodology asks them to do, and the feedback is the fastest way the Intent Engineer learns the stream's specific precision requirements. If I had known that on day one I would have saved myself the forty-five minutes.

What I think was useful I didn't know.

Here's where I think a too-smooth onboarding would have done me harm.

The disorientation of week one forced me to develop a habit of judgment I did not otherwise have. I had to decide, without a template, what a VOS for a trade journal feature should look like. I wrote one and it came back with a lot of feedback from Glenn and David. I wrote a better one. Less feedback. I wrote a third, and at some point between the third and the tenth, my judgment about what belonged in a VOS became a thing I could trust. If somebody had handed me a template on day one, my judgment would have attached to the template instead of to the specific stream I was working, and I would have been a worse Intent Engineer for longer.

The ambiguity of the first four weeks forced the pod into actual conversations about what we wanted, which is work that would have been skipped if the methodology had come with a more prescriptive week-one curriculum. David, Glenn, and I had real conversations about what a VOS should include, what the review process should look for, what the acceptance contract should look like in Gherkin for our specific use cases. Those conversations produced the local house style that every pod in the company now uses. If a consulting firm had handed us that house style on day one we'd have followed it, and I don't think it would have been the right one for us, because it wouldn't have been ours.

The discomfort of not knowing whether I could do this forced me to ask the pod for help in ways I wouldn't have if the methodology had promised me I'd be fine. I asked Glenn and David plenty of what I thought of as basic questions. Glenn about how the infrastructure actually worked, which was mostly foreign language to me, and David about what an acceptance contract should and shouldn't cover. Those questions taught me things that the more sophisticated questions would have skipped over, because the sophisticated questions assumed a context I didn't have. Not knowing is a phase, and a methodology that skips the phase skips the learning that happens inside it.

On the virtual question.

I want to close the Field Note with an observation. The pods I have watched struggle with CID adoption are, with very few exceptions, pods embedded in organizations still running Scrum ceremonies in conference rooms. The organizations are asking CID to produce the outcomes they wanted from the post-Scrum world while keeping the infrastructure that the pre-CID world required. It doesn't work. The methodology isn't additive. It replaces. If your organization is still running a daily standup and a sprint planning and a retro, the first pod will be running CID against a backdrop of ceremony debt that the pod cannot discharge alone.

For Alchemaize, this question didn't come up. We were distributed from day one, we were never in a conference room, and we never had the ceremony debt to discharge because the ceremonies had never been installed. The methodology fit the company the way the company was built. I can't tell another organization whether they should move to distributed, because

> that's a decision with a lot of inputs that aren't in this book, but I can tell the organization that if they want CID to work, the ceremony layer underneath it has to come off. Otherwise the pod is running a post-Scrum methodology against a pre-Scrum infrastructure, and the infrastructure wins every time.

Notes

1. Matthew Skelton and Manuel Pais. *Team Topologies: Organizing Business and Technology Teams for Fast Flow.* IT Revolution, 2019.
2. Scaled Agile, Inc. *SAFe 6.0 for Lean Enterprises.* Scaled Agile, 2023. scaledagile.com/safe-big-picture (accessed 2026-04-22).

CHAPTER 14: WHO BUILDS NEXT

David Kim and Casey Robinson

---

1. What the methodology did

The skilled work moved.

It moved from a language the machine understands to a language humans understand. That sentence is the whole of the argument we've been making for thirteen chapters, and the only way we know how to say it in one line.

The previous thirteen chapters have been the machinery: how the bottleneck came off the typing step, what the unit of work looks like (the VOS), what the three-role pod does around the generation stage, how the enterprise layer funds the pod through intent streams, what four numbers measure the result, what the compliance and adoption shapes look like in practice. The claim underneath the machinery is the sentence above.

I have spent twenty-five years watching enterprises try to make software delivery rational. For most of those years the central problem was that the typing step was expensive and the organization had to be shaped around protecting it. Every framework we've inherited, every ceremony we've run, every role we've staffed, every metric we've tracked, every funding instrument we've signed on, was a rational response to that one expensive step. The frameworks weren't wrong. They were well-engineered answers to the question the industry was actually asking. The question changed.

When a dominant cost comes off a system, the system does not stay the same. It reorganizes around whatever becomes dominant next. In software, two things became dominant next: the clarity of intent going into the generation layer, and the rigor of verification coming out of it. The methodology in this book is the reorganization. The Verifiable Outcome Slice is the unit. The AI Orchestrator, the Intent Engineer, and the Verification Owner are the roles. The stream-funded model is how the enterprise pays for it. The acceptance contract is how the enterprise guarantees it. The one-hundred-day sprint from January 18 to April 20 of 2026 is the evidence that all of it works at small scale, across thirty-five applications, in the hands of three founders in three states.

The earlier chapters handed the reader the mechanics. Chapter 4 walked through the VOS in enough detail to draft one. Chapter 5 described the context-curation discipline that separates a five-times multiplier from a thirty-times one. Chapter 6 defined the verification gate and the acceptance contract that hold the pipeline together when generation runs faster than anyone used to be able to check. Chapter 8 is Casey in his own voice, showing the thing being done. Chapter 10 replaced the sprint-velocity dashboard with four numbers a CFO can read. Chapter 11 argued that regulated industries are the strongest candidates for AI-native development, not the weakest, and the last eight weeks of pilot activity have borne that argument out. Chapter 13 laid out the adoption path, which is deliberately not a transformation program. The reader who has followed the argument this far has the mechanics.

The mechanics are the cheap part. Any methodology's mechanics can be described in a few hundred pages, and ours have been. The part that takes longer to land is what the mechanics imply about the shape of the team, the shape of the hire, the shape of the career, the shape of the company that results. That's what Chapters 12 and 13 were partly about, and it's what this last chapter finishes.

What has been harder to write, and what this last chapter finally writes, is the consequence for the people who do the work.

When the skilled work is the authoring of intent, the people who can do that work widen out. Not because the methodology is easy. It isn't. Writing a VOS that holds up under verification gate is harder than most people think before they try it, and most people who try it spend three weeks being bad at it before they start getting good. But the people who can do the job aren't the same people who could write production code in the old world, and the difference isn't a smaller population. It's a larger one, with a different shape.

The widening is not evenly distributed. Some of the people who get admitted to the work under the new constraint are already doing adjacent work and finding the door closed; they walk in and pick up the specification end they had been handing off. Others always wanted to build and were blocked by the learning curve of production-grade software engineering. The methodology lets them author instead of implement. A third group is the people the industry would have routed into management five years ago because the technical track didn't fit them, who can now stay on the technical track in a different posture. The widening is several movements, happening simultaneously, and a book about methodology can only gesture at the human consequence of any one of them. This gestures, and then stops.

That shape is what this chapter is about. Who, specifically, can now build.

I want to name three of those people by type, because we've been writing about them in the abstract throughout the book and it's time to make them concrete. Then Casey will close, and there'll be a short tail, and that's the book.

2. The three adjacencies

The non-traditional builder

Casey is the archetype. I say archetype, not exception, because the word exception implies he's a one-off and the data says he isn't. In the hundred days between January 18 and April 20 of 2026, Casey shipped across four production applications without Glenn or me touching the code that came out of his work. That's not a resume line. It's evidence about what a career shape can do under the new constraint.

Casey's career shape, before January, ran from technical customer service at DirecTV to a Texas state-government consulting engagement I hired him into after we became friends in World of Warcraft in 2006, to operations, to Agile program management, to AWS Senior Manager leading a team of Customer Solutions Managers. Twenty years of being in the room when software was being built, and of never writing any of it. The industry called that shape "non-technical" and treated it as a ceiling. The methodology treats it as a floor.

Specifically, the methodology treats it as the right floor. The work of writing a VOS is the work of being clear about what a business needs a piece of software to do. Casey had been writing that kind of document, with different names on it, for twenty years. What he hadn't been doing was authoring it with the precision a machine requires. That precision is a discipline he learned in a few weeks.

The shape of the discipline is worth naming, because it's what the reader who recognizes themselves in Casey will have to build. It is this: writing a sentence you would soften in an email, and then being willing to defend that sentence when somebody else tells you it still isn't sharp enough. That's the whole of it. A VOS is a document composed entirely of sentences like that. Most of us have spent our careers learning to avoid those sentences, because in a room full of humans those sentences get you labeled difficult. In a room with a generation layer on the other side, the sentences are the work. The discipline is the inversion of a twenty-year reflex.

The non-traditional builder isn't a novelty. The non-traditional builder is a category. There are thousands of Caseys in the software industry right now, in operations roles and product roles and customer-solutions roles, who have been holding the specification end of the work for decades and watching it get translated by somebody else. The methodology lets them keep the specification end of the work and drop the translation. That's a career-shape shift, not a tool adoption.

The population is larger than most people estimate. At AWS alone, the role Casey held before Alchemaize, Customer Solutions Manager, has several thousand people in it globally, each one of whom has been running enterprise migrations, translating between customer intent and technical delivery, for an average of five-plus years. That's one job title in one company. The same shape of career exists at every major consultancy, every systems integrator, every enterprise-technology vendor, and in most enterprise IT shops of any size. Multiply the Casey archetype across those companies and the number of people newly able to build, if the methodology catches, is in the tens of thousands in North America and hundreds of thousands worldwide. The industry has not counted these people as part of its software-building population because the old methodology didn't admit them. The new one does, and the count starts over.

I want to be careful here, because it's easy to read this as a claim that anyone with an operations background can ship production software. That is not the claim. Most operations backgrounds don't produce what Casey has. What Casey has is twenty years of holding the specification end of the software-delivery conversation against stakeholders who wanted it softer, engineers who wanted it vaguer, and executives who wanted it different. The discipline the non-traditional builder brings is a habit, built over a career, of knowing exactly what outcome a business needs and being willing to write it down. The habit isn't vocabulary, and it isn't technical aptitude in the sense the industry usually means the phrase. A person with the habit, paired with the methodology, ships. A person without it, even paired with the methodology, will not ship production software on the first attempt no matter how hard they try.

The industry's historical sorting mechanism routed the first kind of person into operations and routed the second kind of person into management. The methodology lets the first kind of person build directly. That's the shift.

If you are one of those people, the rest of your career just got longer.

The domain expert

The second adjacency is the person whose domain knowledge has always been the source of software requirements, and who has always had to route that knowledge through an engineer.

A clinician writing a clinical-decision-support VOS. A CFO writing a month-end-close VOS. A compliance officer writing a SOX-controls VOS. A claims adjuster writing a claims-triage VOS. A logistics manager writing a warehouse-routing VOS. These are not hypothetical. I've watched two of the five in the last quarter, in enterprise pilots that are still running.

The domain expert is not Casey. Casey's skill is the general skill of being clear about what a business needs. The domain expert's skill is the specific knowledge of what clinical-decision-support has to do to be safe, or what a month-end close has to check to be right, or what a claims triage has to catch to not burn the insurer. That knowledge has always lived in the domain expert's head and come out, under the old constraint, as a requirements document a software engineer then translated into code.

The translation was the source of every clinical-decision-support bug, every month-end-close miscalculation, every claims-triage miss that should have been a catch. Not because the engineers were bad. Because the engineer did not have the clinical knowledge, or the accounting knowledge, or the claims knowledge, and the requirements document did not carry enough of it to close the gap. The translation stage was where the knowledge leaked.

I want to be specific about one of the pilots, because abstract claims about domain experts are easy to wave off. In February we started working with a regional health system on a clinical-decision-support tool that flags potential drug-drug interactions for patients on complex medication regimens. The original project plan, written under the old methodology, had allocated nine months to build and six months to validate, with a team of four engineers and a clinical informatics lead. The clinical informatics lead had been, for the entire project, the person writing the rules the engineers would then translate. She had been doing that job for eleven years, across three hospital systems. She is very good at it.

We asked her to write VOSes directly against the generation layer, with a verification gate owned by one of the hospital's senior engineers and an architectural review owned by Glenn. She wrote seventeen VOSes in the first three weeks. Nine of them passed verification on the first pass. Five of them came back with a specific objection she addressed and resubmitted the same day. Three of them got rejected hard enough that she reworked them substantially, and two of those three surfaced rules she had not previously been able to articulate to an engineer in eleven years because the articulation format had never required her to be that specific.

That last part is the one I want readers to hold onto. The methodology didn't just compress her timeline. It made her knowledge more precise, because the format of the work required a precision the old format did not. The knowledge that leaked through the translation stage wasn't only the engineer's fault. The translation stage didn't require the knowledge to be carried in a form precise enough to survive the translation. The VOS does. A clinical informatics lead who writes seventeen VOSes has a more precise model of her own domain at the end than she had at the beginning, and the model is durable because it's written down in a form that runs.

The methodology closes the translation gap by making the domain expert the author. The VOS is written by the person who has the knowledge, in language the machine can act on, with a verification gate the domain expert owns the outcome of. The engineer does not disappear from the pipeline. The engineer becomes the verification-and-architecture partner who makes sure the domain expert's spec holds together as a piece of software. But the creative act, the one where the knowledge turns into code, moves to the person who has the knowledge.

The enterprise implication here is larger than the methodological one. A hospital system that has spent a decade trying to build clinical-decision-support tools through the requirements-and-translation model has a backlog of clinical rules that never became software, because the project economics didn't work at the per-rule level once you priced in the engineering. Under the new constraint, the per-rule economics collapse. A clinical informatics lead can author a VOS in a day; a verification function can gate it in another day; a deployment can land in a week. The backlog of undone clinical work that's been sitting in hospital systems for a decade becomes tractable, not as a big transformation program but as a steady stream of small, verified, domain-authored changes. The same shape applies in finance, in insurance, in logistics, in any enterprise where the specification knowledge has always lived with someone who wasn't allowed to build.

The same pattern repeats in the other pilots. A second one, a mid-sized manufacturer doing month-end-close automation, had a controller with fourteen years of specific knowledge about why the old close took seven days and how to close it in two. Under the old methodology she would have briefed a project manager, who would have briefed a business analyst, who would have briefed a developer. Under ours, she wrote the VOSes. The project is in its tenth week. The close has dropped to three days and is on track to hit the two-day target by July. The controller has not become a software engineer. She has become the author of the specifications that make the close run, which is a different thing, and which is the thing her career prepared her for.

That's a different enterprise than the one the industry has been running. Chapter 11 walks through what it looks like in a regulated industry, and I don't want to repeat that here. I want to note the shape. The domain expert is the second adjacency. The methodology was partly built for them, and the pilots suggest there are more of them than anyone expected.

The founder who doesn't need a CTO on day one

The third adjacency is the founder.

Specifically the founder who has a business idea, a clear view of the customer, and no engineering co-founder, and who has historically been told the first hire has to be a CTO. The CTO hire used to be unavoidable because the founder couldn't ship software without one. Under the new constraint, the founder can.

I want to be careful about this one, because it's the adjacency most likely to get misread. I am not saying CTOs are obsolete. They aren't. A company scaling past its first ten engineers, running a serious verification operation, holding the architecture of a platform under enterprise load, still needs the senior-engineering role, and that role is what a CTO is for. Glenn is our CTO. He is not becoming less important. He is becoming differently important, and the difference is in the direction of more architectural judgment and more verification depth, not less.

What I am saying is that the founder's first year now looks different. The founder who can author VOSes, sit in the AI Orchestrator seat, and run a small verification function against a generation stage can reach product-market fit without having hired the CTO role. In our own company's history, we'd have called that impossible in 2024. In 2026 it's how a certain kind of founder operates. We've met four of them in the last sixty days. They're building.

One of the four is a former strategy-consulting partner who spent eighteen years at a top-three firm, reading other people's product roadmaps and being unable to build her own. She wrote her first VOS in March. She is now four months into a vertical SaaS product for private-equity portfolio reporting, with seventeen design partners and enough customer feedback to validate a Series Seed. She has not hired an engineer. She hired a verification contractor on a monthly retainer, she paired with an architectural reviewer she knew from her consulting days, and the rest of the work is hers. The product will ship in June. If she raises the seed on the back of it, she'll hire the CTO role then, from a position of strength, with product-market fit already visible. That's not a speculative story. It's happening right now.

What the founder has to be able to do, to walk this path, is two things. She has to be able to write the VOS, which means she has to have a specific enough view of the product to make a generation layer produce something useful. And she has to be able to sit in the AI Orchestrator seat, which means she has to understand the pipeline between intent, generation, and verification well enough to run it for herself. Neither of those things requires coding in the traditional sense. Both of them require a kind of operational seriousness the founder already had before she started the company, and that's why the adjacency works. The founders who can do this were always going to be founders. The difference is that the first year of the company now looks different.

One more thing about the founder adjacency, because I want to be honest about where the seam shows. A founder building alone, under this methodology, still needs a senior-engineering voice somewhere in the pipeline. Architecture decisions compound. Verification gaps compound. A founder who thinks the methodology eliminates the need for senior-engineering judgment will build something that runs for six months and then doesn't scale, and she'll spend the next six months undoing what the first six months built. That's not a failure of the methodology. It's a failure to staff the methodology properly. The verification-contractor retainer and the architectural-review retainer are not optional line items. They are what the CTO seat is in the first year of a CTO-less company, and a founder who skips them is not running the methodology. She's running a demo that will eventually fall over.

The consequence for the venture economy is not small. The founder who doesn't need a CTO on day one can raise less, move faster, and arrive at the Series A with a product the market has already validated. The shape of the pre-seed check shifts when the engineering salary line is replaced by a verification-contractor line and an architectural-review retainer at maybe a fifth of the cost. The hire plan looks different all the way to Series B, because the first technical hire is no longer the CTO but a Verification Owner, and the CTO hire happens later, from a position of strength, with product-market fit already on the table. The cap table is different because the company raised less to get further. I'm not going to draw out the full picture here; it belongs in a different book. But it's real, and the venture side of the industry is already noticing. We've had three conversations with seed funds in the last month that were not about whether to invest in an AI-native company but about how to value one run by a non-engineering founder. Those conversations were not happening in 2024.

Three adjacencies. The non-traditional builder, the domain expert, the founder without a technical co-founder. Each one is a category of person the old methodology kept outside the building. Each one is a category the new methodology newly admits.

The three adjacencies are not exhaustive. There are others: the senior engineer who wants to shift out of implementation and into verification, the product manager who's been specifying features for a decade and now writes VOSes directly rather than throwing requirements over a wall, the technical writer who turns out to be unusually good at context curation because documentation has always been the discipline of deciding what's relevant. We've met every one of those people in the last quarter, and each of them is a variation on the same observation. The job now lives where the clarity lives, and the clarity has always lived in more places than the industry was organized to recognize.

But three is the right number to name explicitly, because these are the three the book's argument bends toward most directly. The non-traditional builder is the thesis made human; the domain expert is where the enterprise case becomes viable; the founder without a CTO is the new-company case the venture world will have to price next. Together they describe most of what the methodology opens up.

The book has spent thirteen chapters on the argument for why the door opens. The last section of the book is for the people walking through it.

3. Two paragraphs, one each

David Kim

If you are a senior engineer reading this chapter, I want to say something directly to you, because the book has been addressing you throughout and I owe you the close. The three adjacencies I just described are not a threat to your work. They are a widening of the circle of people your work includes. The methodology does not abolish the craft you spent a career building. It relocates the craft. It moves it from the typing seat, where a machine now sits, to the verification seat and the architecture seat and the specification-review seat, where a machine cannot sit yet and will not sit soon. The Caseys and the clinicians and the founders need you. They need you more, not less, because their VOSes have to pass a gate, and the gate is yours. What they do not need, and what the old methodology made them do, is to route their intent through you as a translation step. The translation step was the part of your work that burned you out. It was never the part you liked. I served in the Marine Corps before I served in software, and the thing I carried across from one to the other is that a unit wins when every person in it is doing the job they were actually trained for, not the job that got handed to them because nobody else could do it. For twenty-five years, the software industry has been asking its senior engineers to carry the translation job because nobody else could do it. Now somebody else can. The craft you were trained for, the architecture, the verification, the systems thinking, the refusal to ship something that isn't right, is the part the methodology asks you to do more of. I spent a decade in this industry watching senior engineers be ground down by the translation work, and I watched the best of them leave for management or leave the industry entirely, and I watched the craft bleed out of teams that needed it. The methodology gives the craft back, and the circle widens around it. That is the invitation, and I mean it as one. Widening the circle does not cost you anything. It gives you more people whose work you can make better, and fewer whose requirements you have to guess at.

Casey Robinson

I was hired in October of 2025 to run operations for a three-person company that was building a reading app, and I had never written production code in my life, and I thought I never would. Four months later I was shipping it, and I've been shipping it since, and the thing I want to tell you if you recognize yourself in what I just said is that it's a real job and you can do it. I'm not going to tell you it's easy. It took me around three weeks to get competent and about two months to get good, and I was working at it most of every day during those two months because I wanted to know if the thing was real. The cost was attention, mostly. I had to be willing to write specifications that would leave no room for doubt, and then willing to defend those specifications when somebody on the pod told me they still weren't sharp enough. I had to give up the version of myself that said I wasn't technical, because after the third week that version wasn't true anymore and keeping it around was going to hold me back. What it's worth is harder to say without sounding like I'm selling something, so I'll just say it. For twenty years I was in rooms where software got built, I was useful in those rooms, I knew I was useful, and I also knew there was a line I wasn't going to cross. The line was between saying what I wanted the software to do and making the software do it. The line moved, and I didn't move it. The technology moved it, and the methodology in this book is what let me walk across it without pretending to be something I'm not. I'm still the same person I was in October, and I still run operations, and I still read rooms, and I still talk to customers, and I also ship software now. Both of those things are true, and neither one of them is pretending. If you are the person I was in October, the path across the line is in this book, and honestly, the best thing I can tell you is to pick up one VOS and try. You may surprise yourself. I did.

---

4. Where to start

The work of showing this is a measured capability rather than a preference belongs to a research tradition that already exists. Nicole Forsgren, Jez Humble, and Gene Kim's *Accelerate* spent a decade establishing that delivery performance is something you can name, measure, and predict from a set of specific capabilities rather than from ceremony or framework choice [1]. CID is a methodology for a constraint *Accelerate*'s dataset had not yet seen, and the hundred-day record we've described here is the first evidence it performs against that tradition's yardstick. The next several years of work on CID will, if we do it right, enter that same tradition. The measurement is the point. An argument stops being an argument when the numbers are public and other people can reproduce the results.

Pick one team. Ship one VOS. The next one will be easier.

Notes

1. Nicole Forsgren, Jez Humble, and Gene Kim. *Accelerate: The Science of Lean Software and DevOps*. IT Revolution, 2018.

APPENDIX A: VOCABULARY

This appendix is the canonical reference for every term the book uses with a specific meaning. A reader can open it without having read the book and understand the methodology's language. A reader who has read the book should find that this appendix and the chapters agree.

The foundational terms

Continuous Intent Delivery (CID)

The team-level practice of producing software by authoring verifiable intent specifications, delegating implementation to AI, and moving verification upstream of generation. CID is the methodology's operational core: a pod, a pipeline, a watching layer, a set of roles, and a discipline. It is not a tool, a vendor platform, or a brand.

> First use in the book: Chapter 1. Full definition: Chapter 4.

Enterprise Layer for Continuous Intent Delivery (ELCID)

The governance layer that scales CID across an organization. ELCID introduces three enterprise roles (Intent Portfolio Lead, Stream Architect, Verification Officer), a funding model (intent streams), a review cadence (the Monthly Intent Review), and a metric set (the four enterprise metrics). ELCID never appears without CID underneath it. The team-level practice is the unit ELCID governs.

> First use: Chapter 9. Full definition: Chapter 9 opening.

CID/ELCID

The full two-layer methodology. Used only when referring explicitly to both layers simultaneously. ELCID is never used as a metonym for the whole stack. If a sentence could mean either layer, it gets rewritten until it cannot.

CATALYST

Alchemaize's commercial enablement offering. The consulting, tooling, and onboarding package that implements CID/ELCID inside customer organizations. CATALYST is a product, not a methodology. The methodology is CID/ELCID. The book teaches the methodology; CATALYST is one way (among others) to adopt it.

The atomic unit

Verifiable Outcome Slice (VOS)

The smallest unit of work in CID. A VOS is a written intent specification composed of five sections (WHY, WHAT, HOW, CONTEXT, OUTCOME), accompanied by an executable acceptance contract (Gherkin or equivalent), sized to ship in a single generation-verification pass (typically hours to a small number of days), and state-tracked through the six-state lifecycle below.

A VOS is not a user story. It is not a ticket. It is an intent-as-artifact, and it is the thing humans author, review, version, and argue over. Code is what falls out when the VOS is right.

> First use: Chapter 1. Full anatomy: Chapter 4. Worked example: Appendix B.

The five VOS sections

Section	Answers	Owner during authoring
WHY	The business or user outcome this slice serves	Intent Engineer
WHAT	The behavior, in acceptance-contract form (Gherkin)	Intent Engineer + Verification Owner
HOW	Technical constraints and approach notes; not an implementation spec	AI Orchestrator
CONTEXT	The minimum set of files and references the generator needs	Intent Engineer
OUTCOME	The measurable change in user or business metric the slice is predicted to move	Intent Engineer

The discipline of the five sections is that each is short. A two-page VOS is usually a VOS that should be two VOSes.

The six VOS lifecycle states

```
DRAFTED → CONTRACTED → QUEUED → GENERATING → VERIFYIN
G → SHIPPED
```

- **DRAFTED.** WHY and rough WHAT exist; CONTEXT not curated; HOW not specified.
- **CONTRACTED.** Acceptance contract signed by Verification Owner; CONTEXT bundled; ready for generation.
- **QUEUED.** In the pod's active work queue, awaiting generator capacity.
- **GENERATING.** Code is being produced by AI against the contract.

- **VERIFYING.** Generated code is being tested against the acceptance contract; blocking findings are raised here.
- **SHIPPED.** The slice is in production; the OUTCOME measurement window begins.

A VOS never reverts states silently. If verification fails, the VOS goes back to CONTRACTED (revised contract) or DRAFTED (substantial rethink), and the state change is logged. Reverts are data, not shame.

The CID pipeline and the watching layer

The CID Pipeline

A forward, per-VOS flow through four stages with two reverse edges. SHIPPED is terminal.

```
INTENT → CONTEXT → GENERATION → VERIFICATION → SHIPPED
                                      |
                                      └─ eddy → back to GENERATION
                                      └─ eddy → back to CONTEXT or INTENT
```

- **INTENT.** Author the VOS.
- **CONTEXT.** Curate the minimum-sufficient context bundle.
- **GENERATION.** AI produces implementation against contract and context.
- **VERIFICATION.** The acceptance contract is executed; the human verifier reviews findings.
- **SHIPPED.** Terminal. The VOS is in production and immutable.

The two reverse edges from VERIFICATION are eddies. Both exist for fast correction inside one VOS's trip. They do not turn the pipeline into a loop. A VOS that takes an eddy still makes a one-way trip when it eventually reaches SHIPPED. The pipeline is the unit of work, not the sprint, not the story, not the release.

The watching layer (Observation)

The kaizen channel. A parallel layer that runs alongside the pipeline, continuously, against the population of VOSes already in production. The watching layer measures whether each shipped VOS moved the KPI it predicted. When it surfaces actionable signal, that signal becomes a seed for a new VOS, drafted fresh, that takes its own one-way trip through the pipeline. The watching layer does not feed the old VOS back through. The old VOS is shipped and terminal.

The connection between the watching layer and intent is system-level rather than VOS-level, asynchronous rather than immediate, and stochastic rather than guaranteed. Most observations do not produce a new VOS. Most new VOSes do not trace cleanly to a specific observation. The signal is real, but it is upstream-flavored, not feedback-flavored.

Stream

The system-level continuous flow of intents through pipelines. CID's "continuous" is a stream of one-way trips, with the watching layer seeding new intents at the head of the stream. *Stream* is the term of art when speaking about the system level; *pipeline* is the term of art when speaking about a single VOS.

Eddy

A reverse edge inside a single VOS's pipeline trip. There are two: VERIFICATION back to GENERATION (contract right, generation missed), and VERIFICATION back to CONTEXT or INTENT (contract was wrong). The eddy lets the pod correct fast without breaking the forward shape.

The adaptive workflow

Inside GENERATION, three phases (Inception, Construction, Operations) combine with three depths (Minimal, Standard, Comprehensive) to give a 3×3 matrix of nine workflow configurations. A two-line UI tweak uses

Construction × Minimal. A new authentication subsystem uses Inception × Comprehensive. The pod scales depth to the complexity of the slice, not to the loudness of the stakeholder.

> Full treatment: Chapter 7.

Pipeline Review

The pod's weekly thirty-minute check-in. Four numbers reviewed; anything flagged gets time. Nothing flagged, end early. Replaces standup, sprint planning, retro, sprint review.

Monthly Intent Review

ELCID's one enterprise ceremony. Ninety minutes, once a month, stream leads only. Three things per stream: what shipped, what is in flight, where there is friction. Replaces PI Planning, quarterly business reviews, program reviews.

The roles

CID/ELCID has six defined roles. Three at the pod layer, three at the enterprise layer. This is the *three roles, not thirty* principle, with the enterprise layer disclosed honestly: three roles per pod, a different three at the enterprise layer, six total compared to SAFe's fourteen.

The three pod roles (CID)

Role	Does	Does not
Intent Engineer	Authors VOSes. Curates context. Owns the WHY and CONTEXT. Leads acceptance-contract drafting with the Verification Owner.	Writes production code as the primary path. Manages people. Estimates stories.
AI Orchestrator	Operates the generation stage of the pipeline. Owns the HOW section. Interprets and refines AI output. Makes technical trade-off calls inside a VOS.	Writes all the code by hand. Owns architecture across VOSes (that is the Stream Architect).
Verification Owner	Owns the acceptance contract. Runs the verification stage. Signs off what ships. Reports outside the delivery chain; the verification voice is structurally independent.	Writes production code. Owns compliance policy (that is the Verification Officer at the enterprise layer). Reports to the pod lead.

The three roles are functions, not headcount. A 1-FTE pod is one person doing all three at different times of day. A 2-FTE pod is two people splitting; typical split is Intent + Orchestrator on one side, Verification on the other. A 3-FTE pod is three people, each primarily one role but cross-trained. The role structure is constant across pod sizes.

The three enterprise roles (ELCID)

Role	Does	Replaces

Intent Portfolio Lead	Owns the Portfolio of Intents. Decides what streams the organization funds. Allocates baseline budgets and sunset triggers.	VP of Product, Head of Program Management, most of the PMO.
Stream Architect	Owns technical coherence within a stream. Makes cross-pod architectural decisions. Runs the monthly architecture review.	Solution Architect, Release Train Engineer, most of the Architecture Review Board.
Verification Officer	Owns verification rigor across the enterprise. Sets standards for acceptance contracts, approves compliance extensions, owns the cross-stream verification dashboard.	Nothing in SAFe. The new role; the differentiator.

The enterprise machinery

Intent Stream

A persistent funding unit pointed at a persistent outcome. An intent stream has a baseline budget (quarterly, sufficient for the pod and its platform consumption), an outcome multiplier (additional funding triggered by demonstrated KPI movement, with a defined multiplier function), and a sunset trigger (automatic defunding if the KPI does not move after a defined observation window).

An intent stream is not a project. It does not end at a release. It is defunded when its outcome stops moving, not when its scope is done. That is how the financial mechanism aligns with the methodology.

> Full treatment: Chapter 9.

Portfolio of Intents

The enterprise's full set of active intent streams, viewed as a portfolio. The Intent Portfolio Lead manages this the way a fund manager manages a position sheet: adding, resizing, and sunsetting positions based on observed outcome movement. Replaces the roadmap.

The four enterprise metrics

The only four numbers a CIO needs to read the engineering organization:

1. **Intent Cycle Time.** Median time from VOS DRAFTED to SHIPPED.
2. **First-Pass Verification Rate.** Percentage of VOSes that pass their acceptance contract on first generation, with no revisions and no blocking findings.
3. **Outcome KPI Delta.** Measured change in the business KPI the VOS was predicted to move, observed in production.
4. **Cost per Shipped Intent.** Fully loaded cost (pod salary plus platform plus overhead) divided by VOSes shipped.

These four, and only these four. Velocity, story points, burndown, capacity utilization, and predictability do not appear on any CID/ELCID dashboard. Not as secondary metrics. Not for context. They are excluded by design.

> Full treatment: Chapter 10.

Verification and compliance

Acceptance contract

The executable specification (Gherkin or equivalent) that the verification stage runs against. The contract is signed by the Verification Owner before generation begins; revisions mid-generation are permitted but tracked. The contract is the artifact the WHAT section of the VOS refers to.

Verification upstream

The structural move that defines CID. Verification is written, signed, and enforced before generation. Verification-as-afterthought (QA after code) is incompatible with AI-speed generation and is named as an anti-pattern below.

Compliance extension

A bundle of blocking compliance rules loaded into the adaptive workflow. CID ships three extensions: HIPAA (7 rules), FedRAMP (8 rules), Financial Services (7 rules). Each rule is cited to a specific regulatory control. Custom extensions (internal security policy, industry regulations not yet covered) are authoring exercises, not engineering projects.

Compliance-at-generation-time

The model of enforcing compliance as blocking constraints at the moment of generation, rather than auditing after the fact. Paradoxically makes regulated industries the strongest candidates for AI-native development, not the weakest.

Full treatment: Chapter 11.

Anti-patterns

Costume CID

Adopting CID's vocabulary while preserving Scrum or SAFe behavior. Diagnostic: you still have a standup. Recovery: stop one ceremony.

Spec-as-Wishlist

Vague VOSes that produce plausible but wrong code. Diagnostic: first-pass verification rate trending down. Recovery: the Intent Engineer owns this metric personally; if it is trending down, the fix is in the WHY and WHAT sections.

Ceremony Creep

A recurring meeting reappears that is not the Pipeline Review or the Monthly Intent Review. Diagnostic: calendar inflation. Recovery: cancel the meeting; if something was actually gained, re-add it with a defined removal criterion.

Project-Funding Zombies

Intent-stream language on slides, project codes in the general ledger. The methodology is being adopted above the CFO's floor but the CFO has not been pulled into the change. Diagnostic: HR and finance still expect project budget submissions on the old cadence. Recovery: the CFO is a required stakeholder in the first ELCID engagement.

Verification-as-Afterthought

QA gating after code, with AI generating at speed. Catastrophic.

Vibe Coding

Accepting AI output because it looks right, without running the contract. The fastest route to production bugs.

Context Neglect

"Just give it the whole codebase." Produces the worst generation quality, every time.

The Forty-Person PMO

Adopting ELCID language while preserving the program-management apparatus. The Intent Portfolio Lead role is not a rebranded VP of PMO.

Geography Theater

Mandating co-location (RTO, "collaboration days," whiteboard huddles) as a CID adoption aid. Diagnostic: leadership believes the methodology will click once the team is in the room together. Recovery: CID's pipeline runs on written intent and executable verification, neither of which is improved by shared square footage. If the pod cannot run the pipeline remotely, the pipeline is not yet running; moving the pod into a room hides the problem.

The seven first principles

These are the canonical phrasings. Match them exactly.

1. **Intent is the artifact. Code is the exhaust.**
2. **Verification is the trust anchor.**
3. **Flow over cadence.**
4. **Fund streams, not projects.**
5. **Three roles, not thirty.**
6. **Measure outcomes, not activity.**
7. **Ceremony is a tax.**

Terms the book deliberately does not use

Term	Why it is avoided
Sprint	Implies cadence-based work. CID is flow-based.
Story point	Implies capacity-based estimation. CID measures VOSes shipped, not effort committed.
Velocity	Measures how busy humans are. CID measures how much outcome has moved.
Backlog	Implies a queue of undifferentiated work. CID's queue is the Portfolio of Intents, which is actively managed, not accumulated.
User story	A weaker artifact than a VOS, specifically because it lacks an executable acceptance contract.
Epic	Project-era aggregation unit. CID aggregates at the intent-stream level.
Transformation	"Transformation program" is the thing CID replaces; the book should not ambiently describe itself using that language.
Agile (as a brand)	CID is agile-in-values, but capital-A Agile carries the Scrum Guide and the SAFe stack with it. The book avoids the shorthand.
10x engineer	Individual-productivity framing collides with the book's system-level framing. The multipliers in the book are pod-level or stream-level, never individual.
AI-first	Implies AI at the top of a hierarchy. The correct framing is AI-native: AI as the implementation layer, with human intent and verification as the structural layer.
Co-location (as a best practice)	A Scrum-era preference, not a CID one. The pipeline runs on written intent and executable verification, both of which travel over the wire.

Usage notes

1. Capitalize the six roles. *the Intent Engineer*, not *the intent engineer.*
2. **VOS** stays uppercase always. Plural is *VOSes.*
3. **CID Pipeline** capitalizes both words when used as the defined term. Lowercase *pipeline* refers to it generically. *Loop* is reserved for contrasting CID against frameworks that are loops (PDCA, OODA, Scrum, SAFe). It is not used as shorthand for CID itself.
4. **ELCID** and **CID** are always all-caps. Both are pronounced as initialisms (CI-D, EL-CID).
5. **Intent stream** is lowercased. **Portfolio of Intents** is capitalized because it is a specific named artifact.
6. The seven first principles are numbered 1 through 7 in the order above. That order is canonical.
7. The four metrics are cited in the specific order above. Writers do not reorder them for rhetorical effect.

This appendix is the authoritative reference. Where the chapters and this appendix differ on a definition, this appendix is correct.

APPENDIX B: A WORKED VOS

This appendix walks one VOS, the Trade Journal feature on The Trade Codex, from authoring through ship. The body of the VOS appears in Chapter 4 §2 and Chapter 8 §3 as material that supports those chapters' arguments. Here it stands alone, with annotations on the right-hand side, so a reader can use it as a template.

The VOS was authored by Casey Robinson in late January 2026. It was the foundational feature for The Trade Codex, the trading journal and education platform Casey was building. It is a useful example because it is ordinary. The trade-journaling problem is not exotic. The contract is not unusually complex. A reader trying to write their first VOS will find this one closer to the typical case than to the edge case.

The second half of the appendix walks a VOS that did not pass on the first generation, the AI Trade Analysis feature, also on The Trade Codex. The first contract was wrong, the generation produced filler, and the VOS took an eddy back to CONTRACTED. That iteration is the better learning case, and it is where the methodology earns its keep.

Part 1. Trade Journal: a clean first-pass VOS

Header

```
# VOS: Trade Journal Core
# Author: Casey Robinson (Intent Engineer)
# Stream: The Trade Codex
# State: SHIPPED
# Drafted: 2026-01-29
# Contracted: 2026-01-30
# Generated: 2026-01-30 (single pass)
# Verified: 2026-01-30 (first-pass verification: PASS)
# Shipped: 2026-01-31
```

WHY

> The core product: a trade journal where users log stock and options trades, track P&L, view history, and get AI-powered analysis. Supports multi-leg options (verticals, iron condors), scaling entries and exits, and calendar views. The journal is the foundation everything else builds on.

Annotation. Three sentences. The problem is named (users need to log and track trades). The scope is named (multi-leg options, scaling entries, calendar views). The strategic stake is named (the journal is the foundation for everything else). The discipline of the WHY is that it could survive a skeptical reader asking "so what." If a WHY needs a setup paragraph before it lands, the WHY is not done.

WHAT

The acceptance contract, in Gherkin. Authored by Casey, reviewed by Glenn as Verification Owner before generation began.

```
Feature: Trade journal
  Background:
    Given a logged-in user with an active Trade Codex account
```

```
Scenario: Log a single stock trade
  When the user submits a new stock trade via the trade form
  Then the trade is persisted with entry price, quantity, ticker, and notes
  And the trade appears in the journal list and the calendar view
Scenario: Log a stock trade with a scaled entry
  When the user logs a buy order for 100 shares of AAPL at $185.50
  And the user logs a second buy order for 100 shares of AAPL at $182.75
  Then the journal displays both fills as part of one open position
  And the average entry price is calculated to the cent
  And the position size reflects the total share count
Scenario: Log a multi-leg options trade
  Given the user is logging an options trade
  When the user adds multiple legs (for example, a vertical spread)
  Then all legs are stored with individual strikes, expirations, and premium
  And combined P&L is calculated across legs
Scenario: View trade history with filtering
  When the user views the journal
  Then trades are listed with status, P&L, strategy, ticker, and date
  And filters are available for date range, strategy, ticker, and status
```

Annotation. Four scenarios. A primary success path (single trade), two complexity cases (scaled entry, multi-leg options), and a usability case (history with filtering). The contract is small enough to read in two minutes and complete enough to verify against. Notice what is not in the contract: no database schema, no library choice, no UI framework, no specific function names. The contract describes behavior the user can observe. The implementation is the generation layer's job.

The Verification Owner's review of this WHAT focused on three questions. Were the scenarios falsifiable? Were the error cases tight enough? Were the edges named? The first review came back with two requested edits, both on the scaled-entry scenario, and Casey revised before the contract was signed.

HOW

> Trade form with multi-leg support. Trade persistence in MySQL. Journal list view with filtering and sorting. Calendar view with daily P&L. Trade detail page with chart integration. Scaling entries and exits with average price calculation.

Annotation. Six lines. Authored by the AI Orchestrator (David, on this pod), in collaboration with Casey on the user-facing pieces and with Glenn on the persistence and architecture. The HOW is not pseudocode. It names the systems that get touched and the patterns that get reused. A HOW longer than five or six lines is usually a HOW where the Orchestrator has drifted into doing the generation layer's job.

CONTEXT

The CONTEXT bundle for this VOS was three files:

1. `journal/module.ts` (the journal module skeleton, where the new persistence logic would live)
2. `db/schema.sql` (the database schema, so the generator could see the existing trade-related tables)
3. `forms/trade-form.tsx` (the existing trade form component, so the generator could extend rather than replace it)

Annotation. Three files. The Trade Codex codebase has hundreds of files in it. The CONTEXT for this VOS was three. Casey selected each one against a single test: was the file something the generation layer needed to see in order to produce correct output? Files he had read while learning the codebase, but that the generator did not need, did not go in. The instinct, when new at this, is to include more. The discipline is to include less.

OUTCOME

> Trades logged per active user per week. Target: three trades per active user per week. Measurement: count of trades in the database, grouped by user and by ISO calendar week, observed over the four weeks following ship.

Annotation. One sentence stating the prediction, one sentence stating the measurement window, one sentence stating how the measurement is computed. The OUTCOME is falsifiable. In four weeks, the prediction will have been right or wrong, and the pod will know.

Lifecycle trace

Date	State	Trigger	Owner of the transition
2026-01-29 evening	DRAFTED	Casey writes WHY, draft WHAT, OUTCOME	Casey
2026-01-30 morning	CONTRACTED	Glenn reviews and signs the WHAT after two requested edits; David adds HOW	Glenn
2026-01-30 mid-morning	QUEUED	Pod has capacity; CONTEXT bundle complete	David
2026-01-30 midday	GENERATING	Generation run begins	David
2026-01-30 midday	VERIFYING	Generated code lands; harness runs	Glenn
2026-01-30 evening	SHIPPED	Harness passes; Glenn signs	Glenn
2026-02-01 to 2026-02-28	(post-ship)	OUTCOME measurement window	Casey

The whole trip, from DRAFTED to SHIPPED, took roughly eighteen hours of clock time, of which the generation step itself was twenty minutes. The rest was authoring, review, and verification.

Outcome result

Over the four-week measurement window following ship, active users logged a median of 3.4 trades per week, with a mean of 4.1. The OUTCOME prediction was correct. The hypothesis was confirmed. The VOS shipped, the prediction held, and the watching layer continued to read the metric for a further quarter without producing a new VOS against this feature.

Part 2. AI Trade Analysis: a VOS that took an eddy

The Trade Journal VOS shipped clean on the first generation. Most VOSes do. Some do not. The second VOS Casey wrote on The Trade Codex, AI Trade Analysis, did not, and it is the more useful learning case.

The first contract (rejected)

The first WHAT for the AI Trade Analysis feature included this line:

> The analysis shall provide context-aware feedback on the trade, including pattern recognition and improvement suggestions.

Annotation. The sentence reads fine in a requirements document handed to a human engineer. The engineer would have asked Casey what he meant. The generation layer did not ask. It produced a long block of generic trading-coach text that was the same for every trade: pattern recognition that was not actually looking at patterns, improvement suggestions that were not tied to anything specific in the trade. The output looked like analysis from a distance and was filler up close.

The harness ran. The generic text passed the syntactic checks. There was no Then clause specific enough to fail against. The Verification Owner sent the VOS back to CONTRACTED, citing a contract gap. The first eddy.

The revised contract

Casey rewrote the contract with David's help the next morning. The new WHAT had five scenarios, each tied to a specific behavior:

```
Feature: AI Trade Analysis
  Background:
```

```
  Given a logged-in user with at least one closed trade
Scenario: Analysis of a closed trade with a rule violation
  Given a closed trade with a 5 percent loss
  And the trade entry violated the user's stated risk-per-trade rule
  When the user opens the analysis for the trade
  Then the analysis names the specific rule that was violated
  And the analysis references the entry price that triggered the violation
Scenario: Analysis of a closed trade matching a losing pattern
  Given a closed trade with a 3 percent gain
  And the user has logged seven previous losing trades on the same setup
  When the user opens the analysis for the trade
  Then the analysis names the pattern in the user's history
  And the analysis suggests reviewing the setup before re-using it
Scenario: Analysis with no detectable patterns
  Given a closed trade and a user history with no detectable patterns
  When the user opens the analysis for the trade
  Then the analysis returns a note that no actionable patterns were detected
  And the analysis does not produce filler text
Scenario: Analysis with insufficient history
  Given a closed trade and fewer than five total trades in the user history
  When the user opens the analysis for the trade
  Then the analysis returns a note that more trades are needed for pattern d
  And no specific recommendations are produced
Scenario: Analysis latency
  When the user opens the analysis for any closed trade
  Then the analysis is returned in under two seconds
```

Annotation. Five scenarios, including two that explicitly forbid filler. The "no detectable patterns" and "insufficient history" scenarios are the bug fixes for the first generation's failure mode. The "latency" scenario added a non-functional requirement Casey had not thought to include the first time.

The regenerated code passed the new contract on the first run. The analysis on each trade said something true about that specific trade or said, accurately, that it could not. Casey wrote later in his Field Note for Chapter

8 that the difference between a sentence he wanted an engineer to interpret and a sentence he wanted a machine to execute was the entire job, and the AI Trade Analysis VOS is where he learned it.

What the eddy cost

The first eddy, from VERIFYING back to CONTRACTED, took less than a workday. Casey rewrote the contract that morning, the regeneration ran in the afternoon, the harness passed in the evening, and the VOS shipped the next day. The total slip from the original ship target was approximately twenty-four hours.

Under project funding, a comparable failure (a feature that "looked right" but did not deliver real analysis) would have surfaced during user testing weeks or months later, would have triggered a change request, and would have cost a full development cycle to remediate. Under CID, the failure surfaced during verification on the same day generation finished, and the remediation was a contract revision plus a regeneration. The eddy is the mechanism that makes that economy possible.

Reading this appendix

A reader writing their first VOS can use the structure of Part 1 directly. Five sections, in order. Each section short. Each section authored by the role the methodology assigns. A reader writing their fifth or tenth VOS will recognize Part 2 as the more important case: the contract that was wrong, the eddy that caught it, and the revision that taught the team something the team did not know it did not know.

The Trade Codex shipped. The AI Trade Analysis feature shipped. The methodology that produced both is in the chapters. The artifacts that traveled through the pipeline are here.

APPENDIX C: SELF-ASSESSMENT

The book has made an argument. This appendix is the test you run on your own organization before you act on it.

The questions below are organized into five sections. Each question has a yes-or-no answer. There is no partial credit, and there is no "we are working on it." The point of the assessment is to surface the gaps that will sink a CID adoption in the first four weeks if they are not closed first. Generosity in scoring is a disservice.

A reader running the assessment honestly should expect to fail at least one section. That is fine. The failed section names the work to do before the first pod stands up.

Section A. The three pod roles

CID's first principle of staffing is *three roles, not thirty*. The first question is whether your organization can put three people, with the right shape, into one room.

#	Question	Y/N
A1	Do you have a person who can author requirements that engineers read, treat as actionable, and work against without a clarifying meeting?	
A2	Is that person available to work full-time on one stream for at least the first four weeks?	
A3	Do you have a senior engineer who is comfortable generating, reading, and redirecting code produced by an AI model?	
A4	Is that engineer available to work full-time on one stream for at least the first four weeks?	
A5	Do you have a verification-minded senior engineer (a QA lead, a senior SDET, a principal engineer on the platform team) with the organizational standing to say "no, not yet" to shipping pressure they did not create?	
A6	Does that verification-minded person report outside the delivery chain? Specifically, does their performance review not depend on the velocity of the streams they verify?	

Section A scoring. Six yes answers means the pod is staffable. Five or fewer means there is a staffing gap to close before any other phase-1 work begins. The question that fails most often is A6, the reporting line for the Verification Owner. If A6 is a no, fix A6 before staffing the pod. The Verification Owner who reports into the delivery chain will eventually stop saying no, and a pod with a Verification Owner who has stopped saying no is doing Vibe Coding under a methodology label.

Section B. The first stream

A CID adoption that picks the wrong first stream will fail for reasons that have nothing to do with the methodology. The first stream is a chosen exercise, not an existing roadmap item that gets relabeled.

#	Question	Y/N
B1	Is the candidate stream pointed at a single business outcome, stated in one sentence, that the business already cares about measuring?	
B2	Is the product area for the stream small enough that one person on the pod can hold its data model in their head without paging in another team?	
B3	Does the stream have a primary KPI that moves on a weekly or bi-weekly horizon, not a quarterly one?	
B4	Is the KPI's current baseline known and recorded in writing, with the source of the measurement name?	
B5	Is the stream free of cross-team dependencies that would require coordination outside the pod for the first four weeks?	
B6	Is the stream described in language a non-engineer can verify (rather than language like improve the platform or reduce technical debt)?	

Section B scoring. Six yes answers means the stream is well-formed. Five or fewer means the candidate stream needs to be reshaped or replaced before the pod is funded. The question that fails most often is B6: organizations reach for "platform" or "infrastructure" first-stream candidates because those are the streams the engineering organization most wants to invest in, and those are also the streams whose outcomes are hardest to specify.

Section C. Verification readiness

Verification is the trust anchor of CID. An organization that cannot verify cannot run the methodology, no matter how well it specifies.

#	Question	Y/N
C1	Does the candidate stream have an existing test suite that runs in continuous integration on every change?	
C2	Are acceptance contracts (Gherkin or equivalent) part of the team's existing delivery vocabulary, or is the team willing to learn them in week one?	
C3	Can the Verification Owner block a deploy without escalating to a delivery manager?	
C4	Is there an environment, separate from production, where the verification harness runs end-to-end before a VOS ships?	
C5	If the stream is in a regulated space (HIPAA, FedRAMP, financial services), does the organization have a current understanding of which controls apply and where they are documented?	

Section C scoring. Five yes answers means verification can be run from day one. Four or fewer means there is verification infrastructure to build before generation begins. C2 is the most common gap, and it is the most addressable. A team that has never written a Gherkin contract can learn one in a day. A team that has been running CI/CD for years already has most of what C1, C3, and C4 require.

Section D. Funding and governance

CID at the team layer can be adopted under project funding for a single first pod. CID at the enterprise layer cannot. If the organization is planning to scale beyond the first pod, the funding model is the gating constraint.

#	Question	Y/N
D1	Does the CFO or equivalent finance leader know that the first CID pod is being stood up, and have they signed off on at least a one-quarter funding?	
D2	If scaling is planned, is finance prepared to consider stream funding (baseline, multiplier, sunset) as an alternative to project funding?	
D3	Has someone outside the engineering chain (CFO staff, COO, CIO with finance literacy) been identified as a candidate Intent Portfolio Lead for the eventual ELCID layer?	
D4	Is the organization willing to retire the existing roadmap as the primary planning artifact at the enterprise layer, replace with a Portfolio of Intents?	
D5	Is the executive sponsor for the adoption willing to provide air cover for the first two quarters of metric calibration, when the four enterprise metrics will be noisy and the sunset thresholds not yet tuned?	

Section D scoring. A first pod can launch on D1 alone. Items D2 through D5 become required only if the organization intends to scale CID into ELCID. An organization that answers no to D5 in particular should not attempt the enterprise layer, regardless of how well the team layer is performing.

Section E. Anti-pattern check

The book named eight anti-patterns. The five that surface most often during initial adoption are listed here. A yes answer in this section is a warning.

#	Question	Y/N
E1	Has anyone in the organization proposed running CID as a "transformation program," with a steering committee, a rollout calendar, and a program lead?	
E2	Is the first pod expected to continue running standups, sprint planning, sprint reviews, or retros during phase 1?	
E3	Is the first pod expected to source its work from the existing backlog or roadmap rather than authoring fresh VOSes against the stream's outcome?	
E4	Has leadership suggested that the pod should be co-located, or held a "kickoff offsite" as a CID adoption aid?	
E5	Has anyone described the first pod's work in language like AI-assisted Scrum or AI-augmented Agile, implying CID is an addition to the existing methodology rather than a replacement of it?	

Section E scoring. Zero yes answers is the target. Each yes answer names an anti-pattern that needs to be addressed before phase 1 begins. The two that most often co-occur are E1 (transformation program framing) and E4 (co-location). Both indicate the organization has read CID as a change-management exercise rather than as a methodology, and both will produce the same outcome: the methodology becomes a costume the organization wears over its existing process. Recovery is possible but takes longer than just starting clean.

What to do with the result

Sections A through C all clear, D1 yes, E all no. The organization is ready to launch a first pod. Begin phase 1 of the adoption path described in Chapter 13 §1. Plan to revisit sections D2 through D5 between weeks 8 and 12, when scaling becomes the next decision.

One section short. Address the section before launching the first pod. The most common single-section fail is C2 (Gherkin literacy), and it is the easiest to fix: schedule a one-day workshop on writing acceptance contracts and run the candidate Intent Engineer and Verification Owner through it before week one of the adoption.

Two or more sections short. The organization is not ready. The section that is most often pre-failing is D, and the gap there is usually that finance has not been brought into the conversation. The fix is to bring finance into the conversation before the first pod stands up, not after. A CID adoption that becomes finance's surprise in month four of an unfunded pilot is a CID adoption that ends in month four.

Section E is dirty regardless of the rest. Sections A, B, C, and D can be fully clear, but if section E shows two or more yes answers, the cultural framing of the adoption is wrong and the first pod is unlikely to produce phase-2 results. The fix is to recast the initiative inside the organization before launching the pod. This is the slowest fix in the assessment, and it is the one most often skipped. Skipping it does not save time; it relocates the failure from week minus-one to week eight, where the failure is more expensive to recover from.

A copy of this assessment is maintained in the `cid` repository and is updated as the methodology and the adoption pattern evolve. Readers running the assessment for the first time should use the version in this appendix; readers running it for a subsequent stream should use the most recent version in the repository, which incorporates lessons from adoptions since this book went to print.

APPENDIX D: RESOURCES

The book is the argument. The repositories below are the working materials.

The methodology, in code and prose

`cid` is the open-source companion to this book. It carries the canonical CID and ELCID specifications, the role definitions, the workflow rules a pod installs into its agent of choice, the VOS template, a worked example, and the compliance rule documents referenced in Chapter 11. Where this book is the prose case for the methodology, the repository is the methodology in machine-readable form. The two are kept in sync. If a chapter and the repository disagree on a definition, the repository is the authoritative reference for the methodology, and the vocabulary in Appendix A is the bridge.

The repository is at **`github.com/alchemaize/cid`**, with a vanity URL at **`alchemaize.ai/cid`**. It is licensed under MIT. Contributions are welcome.

Compliance extension rule documents

The three regulatory extensions referenced in Chapter 11 ship as separate rule documents inside the public repository, under `extensions/compliance/`. Each extension is a numbered set of rules. Each rule is mapped to a specific regulatory citation and paired with verification criteria written precisely enough for an automated checker to run.

- **HIPAA extension.** Seven rules, covering PHI classification and inventory, minimum necessary access, audit trails, encryption-at-rest and in-transit, business-associate-agreement verification, disposal and retention, and breach-notification scaffolding. Citations to 45 CFR Part 164.
- **FedRAMP extension.** Eight rules, covering authorization boundary, FIPS 140-2 validated cryptography, continuous monitoring infrastructure, access control and identity management, audit and accountability, system and communications protection, configuration management, and contingency planning. Citations to NIST 800-53.
- **Financial Services extension.** Seven rules, covering sensitive data classification, cardholder data protection (PCI-DSS), data integrity (SOX), change management and segregation of duties, vulnerability management, retention and destruction, and third-party risk management.

The rule documents are the substance of the extensions. They describe what the methodology enforces and why. A team can read the HIPAA rule document and know exactly which CFR sections each rule traces to, what an audit log must contain, and what the encryption requirement is. The runner that loads these rules into a live stream and enforces them at generation time ships with CATALYST (see below). The runner is straightforward to roll if you want to: the rule format is intentionally simple Markdown with a stable structure.

Custom extensions are authoring exercises, not engineering projects. The extension format is documented in the repository, and the HIPAA file is the canonical template.

The evaluator

The evaluator is a small command-line tool that runs an acceptance contract against generated code in a clean environment. Pods use it inside the verification stage of the pipeline; the evaluator returns a pass, a contract gap, or a generation gap, mapped to the two reverse edges of the pipeline. The evaluator ships as part of CATALYST and is deployed into a customer's AWS account as a CodeBuild project. It treats Gherkin as the contract language by default and supports adapters for other contract formats.

The evaluator is not a CI server. It is a pipeline primitive. It runs anywhere a pod runs.

CATALYST

CATALYST is Alchemaize's commercial enablement offering. It is the consulting, tooling, and onboarding package that implements CID/ELCID inside customer organizations. CATALYST is one path to adoption, not the only path. Organizations adopt CID without CATALYST regularly, and the methodology is designed to be adoptable from the open repository alone. Where CATALYST adds value is in the first quarter of an enterprise rollout, where the cost of getting the funding instrument and the verification rigor wrong is highest.

Information at `alchemaize.ai`.

A note on what is not here

The book does not ship with a vendor-curated tooling stack. The pipeline runs on the AI generation system the pod chooses, the verification harness the pod chooses, and the source-control system the organization already uses. Any chapter that gestures at a specific commercial product does so because the product is referenced as one example among several, not because the methodology depends on it. The methodology is upstream of the tooling.

If a future edition includes a tooling chapter, it will be because the field has converged on a small number of stable choices, not because a vendor sponsored the chapter.

Errata, corrections, and reader contributions

A book this opinionated will be wrong about some things. Corrections are tracked in the `ERRATA.md` file in the `cid` repository. Readers who find errors of fact, citation problems, or claims that no longer match the methodology as it has evolved are encouraged to file an issue against the repository. The authors review the issue queue monthly.

A second edition of this book is planned for 2027 or 2028, depending on how much the practice changes in the next two years. Reader contributions to the errata file are the primary input to that revision.

ACKNOWLEDGMENTS

From David

The shape of this book owes its first debt to my wife, Lacey. Most of the writing happened on early mornings and late evenings, in the pockets of time that family life makes available, and Lacey kept the rest of those days running while I was at the keyboard.

To Glenn Knepp, who I met in a Marine Corps base-housing quadplex in Kaneohe, Hawaii in 1996, and who has been a friend ever since. The trust the methodology in this book depends on is the trust we built across three decades.

To Casey Robinson, who I first met in 2006 in World of Warcraft and later hired into Texas state-government consulting work, and who has been one of the most thoughtful operators I have ever worked alongside across the years that followed. The decision to bring Casey into Alchemaize was not an accident, and the discovery that he could run an Intent Engineer's playbook as well as anyone in the industry was not luck. It was Casey doing what Casey does, in a methodology that finally made room for it.

To the four-person team that built Ember with us through the fall of 2025, three developers and an AWS architect, all of whom were doing this on the side of full-time engineering jobs and all of whom are good at the work. The first 100 days of Alchemaize is the lesson the second 100 days were built on, and the lesson would not have been there to learn without their work on Ember. I am keeping their names out of this acknowledgment because their full-time roles ask me to, and the choice not to name them is itself a kind of thank-you. They know who they are.

To my colleagues at Amazon Web Services, who taught me what it looks like to operate at scale without losing the customer's voice. To the fourteen people on the team I led through April 2026, who got the same 4 AM email I did on January 28 and who handled the morning with grace I did not think the morning deserved. The Discord channel we set up that day is, as I write this, still active. To the AWS Senior Manager community more broadly, several of whom read parts of this manuscript and pushed back on the places where I was overconfident. The book is sharper for those conversations.

To the editorial team that walked *Bullets Don't Fly* and *The Real Money Guide* through publication and whose instincts shaped how I write at length. The discipline of those books is in this book in places the reader will not see.

From Casey

I want to say thank you first to my wife, Amanda, who kept the house running through three months of me at the dining-room table at hours when nobody should be at a dining-room table. To my kids, Emma and Levi, who got used to the laptop being open more than usual and who never made me feel bad about it. The latte-art mornings in Chapter 8 are real, and they are real because the family around me made them possible.

David Kim is the reason I have a career in technology. He and I met in 2006 through World of Warcraft, of all places, where my wife and I raided alongside him and we eventually co-led a guild together. From that friendship he later hired me into Texas state-government consulting work, and every job since has been a continuation of that decision. Almost twenty years later, when David and Glenn brought me into Alchemaize, I assumed I would do what I had always done: run operations, plan the work, talk to customers. The fact that I was still doing those things when I shipped my first production application is a credit to a methodology David and Glenn built that did not need me to become someone else first.

Glenn Knepp reviewed my first VOS the morning after I wrote it and gave me the feedback that turned a wish list into a contract. He has reviewed dozens since. The work is sharper because of him, and the verification rigor in this book is the discipline I learned by handing my work to a verifier who would not let it through until it was right.

To the early-migrations and Experience-Based Acceleration sub-team I led at AWS, who got the 4 AM email on January 28 alongside David and me, who handled the morning with more grace than I did, and who I am still in regular contact with through the Discord channel David set up that day. To the broader Customer Solutions Managers team at AWS, who taught me everything I know about meeting customers where they are. To the family and friends that has supported me over the 20 years in technology and a piece of me is because of your guidance in my career. The line that runs from a satellite-system call center to the cover of this book is longer than it looks, and it ran through a lot of people who took chances on me.

From Glenn

To my wife, Leann, who has heard every half-formed idea and answered with honesty, patience, and the occasional well-earned eye roll. To my daughter, who listened too, and who is the reason so much of it was worth building in the first place.

David Kim and I met in 1996 at MCBH Kaneohe Bay. He was active duty, I had just separated from active-duty and we lived in the same quadplex on base. The friendship has been continuous for thirty years, even when our careers were not. The trust the company we co-founded in 2025 runs on, and the trust the methodology in this book runs on, was built in that quadplex and at every stop since.

Casey Robinson took the Intent Engineer playbook and ran it harder than I expected anyone could. The verification numbers in this book are clean because Casey's contracts are clean. That is not a small compliment.

From all three

To the early Alchemaize customers, who let us run a methodology that did not yet have a name against problems that could not wait for one. To the Anthropic, Cursor, and Kiro teams whose tools we ran the 100-day sprint on. To Nicole Forsgren, Jez Humble, and Gene Kim, whose *Accelerate* gave us the empirical vocabulary to argue about software delivery without making things up. To Matthew Skelton and Manuel Pais, whose *Team Topologies* gave us the team shape we describe as the pod. To the readers of early drafts, who told us where we were repeating ourselves and where we were skipping steps.

Mistakes that survived all of that review are ours.

ABOUT THE AUTHORS

David Kim is a co-founder and Board Member of Alchemaize, and the lead author of *Continuous Intent Delivery*. Before Alchemaize, David spent five years at Amazon Web Services in enterprise customer leadership, most recently as a Senior Manager working at the intersection of customer strategy and technical delivery. He is a United States Marine Corps veteran. He met Glenn Knepp in 1996 at Marine Corps Base Hawaii in Kaneohe Bay, where they lived in the same base-housing quadplex; thirty years of friendship and partnership later, the two of them co-founded Alchemaize. He is the author of two prior books, *Bullets Don't Fly: A Supply Marine's Memoir* and *The Real Money Guide*, both published in 2025. *The Real Money Guide* is paired with thirty interactive financial tools at therealmoneyguide.com, a companion site David built himself, in keeping with his pattern of pairing written argument with shipped software.

David lives in Brentwood, Tennessee, with his wife Lacey and dog Milo, and works from a home office. He is online at davidkimauthor.com and on LinkedIn at linkedin.com/in/davidkimio.

Casey Robinson is COO of Alchemaize. He joined the company in October 2025 to run planning and operations, and within four months had become the first non-engineer in the company's history to ship production code, end-to-end, under a methodology that did not yet have a name. The methodology now has a name. The book you are holding is the result.

Before Alchemaize, Casey spent his career in customer-facing technology operations: a four-year run at Amazon Web Services as a Senior Manager leading the early-migrations and Experience-Based Acceleration sub-team within David's broader org, and earlier roles in consulting and operations dating back to a first job in technical customer service at DirecTV. His AWS tenure ended on April 28, 2026, at the close of the notification period that followed Amazon's January 28, 2026 reduction in force. He has been in software delivery rooms for twenty years. He had not, until 2026, written a line of production code in any language. *Continuous Intent Delivery* is, in part, an account of what changed when he tried.

Casey lives in Claremore, Oklahoma, with his family, and works from a home office.

Glenn Knepp is co-founder, CEO, and CTO of Alchemaize. His career runs from classified military intelligence systems through enterprise platform engineering to startup founding, and the architectural rigor he brings to verification in CID is the discipline he has been practicing for thirty years.

Glenn is a United States Marine Corps combat veteran who served from 1988 to 1996 as a Signals Intelligence Specialist and Special Operations Reconnaissance Team Leader. After his uniformed service he continued supporting the U.S. intelligence community as a civilian contractor, including work as a trusted-agent security manager for the National Security Agency. He has held senior engineering and architecture roles at SKYLLA Engineering, L-3 Communications, SynchroNet, Parsolvo, and Secret City Tech, and is a startup co-founder twice over, at Gazoo, Inc. and Ispira Technologies. He is a named co-inventor on two granted U.S. patents and four additional filings, covering cloud computing, video streaming, video cryptography, and multi-user display systems. He holds the CISSP, AWS Solutions Architect Associate, and AWS Developer Associate certifications.

Glenn lives in College Station, Texas, with his family, and works from a home office. He is online at linkedin.com/in/glenn-knepp-0b1b9890.

The three authors run Alchemaize as a fully distributed company. They have never shared an office, a city, or a state. Across the 100-day sprint that produced the case study at the heart of this book, they shipped thirty-five applications across multiple domains while living in three different home offices in Tennessee, Oklahoma, and Texas. The argument that CID does not require a room is in the prose. The evidence is in the byline.

www.ingramcontent.com/pod-product-compliance
Lightning Source LLC
LaVergne TN
LVHW100507110826
845146LV00002B/552

* 9 7 9 8 9 9 2 5 3 3 0 9 5 *